GREEN
INFRASTRUCTURE

Map and Plan the Natural World with GIS

KAREN E. FIREHOCK • R. ANDREW WALKER

Esri Press
REDLANDS | CALIFORNIA

Esri Press, 380 New York Street, Redlands, California 92373-8100
Copyright 2019 Esri
All rights reserved
23 22 21 20 19   1 2 3 4 5 6 7 8 9 10
Printed in the United States of America

Library of Congress Control Number:2019942188

The information contained in this document is the exclusive property of Esri unless otherwise noted. This work is protected under United States copyright law and the copyright laws of the given countries of origin and applicable international laws, treaties, and/or conventions. No part of this work may be reproduced or transmitted in any form or by any means, electronic or mechanical, including photocopying or recording, or by any information storage or retrieval system, except as expressly permitted in writing by Esri. All requests should be sent to Attention: Contracts and Legal Services Manager, Esri, 380 New York Street, Redlands, California 92373-8100, USA.

The information contained in this document is subject to change without notice.

US Government Restricted/Limited Rights: Any software, documentation, and/or data delivered hereunder is subject to the terms of the License Agreement. The commercial license rights in the License Agreement strictly govern Licensee's use, reproduction, or disclosure of the software, data, and documentation. In no event shall the US Government acquire greater than RESTRICTED/LIMITED RIGHTS. At a minimum, use, duplication, or disclosure by the US Government is subject to restrictions as set forth in FAR §52.227-14 Alternates I, II, and III (DEC 2007); FAR §52.227-19(b) (DEC 2007) and/or FAR §12.211/12.212 (Commercial Technical Data/Computer Software); and DFARS §252.227-7015 (DEC 2011) (Technical Data – Commercial Items) and/or DFARS §227.7202 (Commercial Computer Software and Commercial Computer Software Documentation), as applicable. Contractor/Manufacturer is Esri, 380 New York Street, Redlands, CA 92373-8100, USA.

@esri.com, 3D Analyst, ACORN, Address Coder, ADF, AML, ArcAtlas, ArcCAD, ArcCatalog, ArcCOGO, ArcData, ArcDoc, ArcEdit, ArcEditor, ArcEurope, ArcExplorer, ArcExpress, ArcGIS, arcgis.com, ArcGlobe, ArcGrid, ArcIMS, ARC/INFO, ArcInfo, ArcInfo Librarian, ArcLessons, ArcLocation, ArcLogistics, ArcMap, ArcNetwork, *ArcNews*, ArcObjects, ArcOpen, ArcPad, ArcPlot, ArcPress, ArcPy, ArcReader, ArcScan, ArcScene, ArcSchool, ArcScripts, ArcSDE, ArcSdl, ArcSketch, ArcStorm, ArcSurvey, ArcTIN, ArcToolbox, ArcTools, ArcUSA, *ArcUser*, ArcView, ArcVoyager, *ArcWatch*, ArcWeb, ArcWorld, ArcXML, Atlas GIS, AtlasWare, Avenue, BAO, Business Analyst, Business Analyst Online, BusinessMAP, CityEngine, CommunityInfo, Database Integrator, DBI Kit, EDN, Esri, esri.com, Esri—Team GIS, Esri—*The GIS Company*, Esri—The GIS People, Esri—The GIS Software Leader, FormEdit, GeoCollector, Geographic Design System, Geography Matters, Geography Network, geographynetwork.com, Geoloqi, Geotrigger, GIS by Esri, gis.com, GISData Server, GIS Day, gisday.com, GIS for Everyone, JTX, MapIt, Maplex, MapObjects, MapStudio, ModelBuilder, MOLE, MPS—Atlas, PLTS, Rent-a-Tech, SDE, SML, Sourcebook•America, SpatiaLABS, Spatial Database Engine, StreetMap, Tapestry, the ARC/INFO logo, the ArcGIS Explorer logo, the ArcGIS logo, the ArcPad logo, the Esri globe logo, the Esri Press logo, The Geographic Advantage, The Geographic Approach, the GIS Day logo, the MapIt logo, The World's Leading Desktop GIS, *Water Writes*, and Your Personal Geographic Information System are trademarks, service marks, or registered marks of Esri in the United States, the European Community, or certain other jurisdictions. CityEngine is a registered trademark of Procedural AG and is distributed under license by Esri. Other companies and products or services mentioned herein may be trademarks, service marks, or registered marks of their respective mark owners.

Ask for Esri Press titles at your local bookstore or order by calling 1-800-447-9778. You can also shop online at www.esri.com/esripress. Outside the United States, contact your local Esri distributor or shop online at eurospanbookstore.com/esri.

Esri Press titles are distributed to the trade by the following:

In North America:
Ingram Publisher Services
Toll-free telephone: 800-648-3104
Toll-free fax: 800-838-1149
E-mail: customerservice@ingrampublisherservices.com

In the United Kingdom, Europe, the Middle East and Africa, Asia, and Australia:
Eurospan Group
3 Henrietta Street
London WC2E 8LU
United Kingdom
Telephone 44(0) 1767 604972
Fax: 44(0) 1767 6016-40
E-mail:eurospan@turpin-distribution.com

Contents

Preface — vii
About the authors — xi

Chapter 1
Green infrastructure: Considering nature before and during development — 1

Chapter 2
Modeling the natural landscape — 23

Chapter 3
The six-step process — 55

Chapter 4
Getting the right data — 81

Chapter 5
Making asset maps — 103

Chapter 6
Assessing risks to your assets — 141

Chapter 7
Determining opportunities — 173

Chapter 8
Implementing GI plans — 203

Appendix A
A brief history of landscape connectivity theory and modeling 231

Appendix B
Technical appendix: Core attributes 239

Appendix C
Esri® Story Maps and other resources 249

Bibliography 251
Index 259

Preface

Esri founders Jack and Laura Dangermond have spent almost a half century developing tools that can help plan and design in ways that protect and connect our environment. Now, Esri has taken on a new challenge—to help map and evaluate the best natural landscapes in the US and beyond through a new green infrastructure (GI) model. As Jack has explained, GI encompasses all the natural resources we need to create livable and sustainable communities. Our natural resources can be thought of as our assets. Just as we manage our communities' built assets such as roads or buildings, we must also manage our natural assets; to do this, we need to determine their extent, their condition, and whether they need protection or restoration. This book helps fulfill Jack's vision of an ecologically intact and healthy landscape that spans the country. These goals harken to early planners and landscape architects, such as Ian McHarg, who promoted the idea that we can design with nature, and Frederick Law Olmsted, who designed parks that linked across landscapes. To realize the vision of a connected landscape requires good maps—and good maps are built with good tools. Esri has partnered with the nonprofit Green Infrastructure Center Inc. to build a national model to enable communities to create GI maps and plans to shape their landscapes in ways that build resilient, sustainable communities for both people and nature.

Fortunately, Esri did not have to model the nation's GI from the beginning—existing models and code created for states by the Green Infrastructure Center could be scaled up to the national level. Yet there were technical challenges on how to run the model for the entire country. Modelers had to identify surrogate data and other datasets to provide necessary information about key habitats and connected landscapes available at a national scale. Esri, together with the Green Infrastructure Center, convened experts in the field at its Redlands, California, headquarters to help address these challenges. The experts included ecologists, modelers, and practitioners of GI planning, who advised the staff at Esri's Applications Prototype Lab. The result was a national model and map output of the nation's best habitats, as well as the necessary apps to customize them. The national model and map were launched at the Esri User Conference in June 2016. This book follows on that effort—providing the instructions and tips for applying the model to regional, state, and local planning needs.

The national model works best at the large-landscape scale, where it can identify, classify, and measure intact core landscapes and the corridors between them and provide an initial ranking of those habitats from the best to the least effective. This model is the first time that a national GI framework has been made available to help connect landscapes, support native species and ecological processes, protect natural cultural resources such as historical sites and outdoor recreation, and promote cooperation among landowners.

Although the national map offers a comprehensive overview of GI landscapes in the US, its most effective application is at the local level—anywhere from the region or state to individual counties. The scalability of the map—from seeing the entire country at once, to drilling down to a state or county park and seeing how they are connected—is the true power of this software.

The data provided by this model can be altered to reflect local priorities, allowing the planner or user to sort information according to which aspect of the planning geography is most important. As a result, you can customize the maps to meet local needs and priorities. For example, you can focus on data by watershed, county, ecoregion, and so on. Just download the data to a desktop environment and add additional map layers, such as known wetlands, or reprioritize a core based on locally rare species or the migration routes of large mammals, such as elk.

For local planning efforts, communities can participate in ranking habitat areas. Communities are encouraged to prioritize GI information and add their own local knowledge to the map. For example, they might overlay other key data that adds priority focus to certain areas, such as their favorite scenic views or sites of special scientific interest. The ability to add details at the local level and to integrate them into county, state, and regional maps and plans demonstrates the incredible potential of this GI model for planners, conservationists, landscape architects, local leaders, and geographic information system (GIS) analysts.

The greatest difference can be made by adding local priorities, goals, and relevant data to local planning. A map is a tool; it provides information that can reveal new understandings, identify areas of need, depict trends, and enable priorities to be set. This book shows users how to apply the national model to craft locally relevant conservation plans for a community.

Audience and format

This book is intended for GIS modelers and analysts, land planners, landscape designers, conservationists and ecologists, land trust managers, professors, and anyone who wants a tool to plan for more effective conservation. It explains how to use the GI model for the US to craft accurate, data-driven maps that facilitate local GI planning on the ground, where land-use decisions are made every day. It provides examples of how to use the US data to create custom maps to meet local needs. The book also has methods you can employ to create maps in areas that may not yet have a model.

The book serves as a technical reference tool. It builds on prior publications by Karen Firehock. For those who are interested in statistics about the value of GI planning, who need help organizing their communities to create a GI plans, or for more examples of completed plans, please see Firehock's prior book, *Strategic Green Infrastructure Planning: A Multi-Scale Approach*.

Chapter 1 of this volume summarizes the case for this work, reviews the history of this approach, and (in Appendix A) lays out the foundational history of the science supporting this approach. Chapter 2 provides an overview of GI modeling and the evolution of that field. Chapter 3 covers the basic process for crafting a GI plan, goal setting, and what to map and why. Chapter 4 explains how to find and apply the right data. Chapter 5 describes the process for creating customized maps and working at the right scales for decision making. Chapter 6 walks the designer through the steps of how to evaluate which resources are at risk; and chapter 7 shows how to use the data of highest-value landscapes and the risk assessment to set priorities. Finally, chapter 8 takes the designer through implementation—creating maps that tell the right story and using them to drive actions for creating healthful and sustainable communities. Appendix A provides a brief history of landscape connectivity theory and modeling. Appendix B covers technical terms. Appendix C provides information about story maps that illustrate concepts covered in this book.

Acknowledgments

Thanks to key leaders in the field mentioned in chapter 1, who provide the foundation for this work; to Jack and Laura Dangermond, whose vision for a connected, resilient landscape made this book and the national model possible; and to the dedicated staff at the Esri Applications Prototype Lab and other Esri departments who spent countless hours adapting the GI model to run for the US. We also acknowledge landscape architect and planner Arancha Muñoz-Criado, who served as Esri's inspiration throughout this endeavor and who developed many of the project's promotional materials and guides.

We recognize and thank several others at Esri: Jennifer Bell, Mark Deaton, Thomas Emge, Witold Fraczek, Bob Gerlt, John Grayson, Jim Herries, Dave Johnson, Hugh Keegan, Ryan Perkl, Mark Smith, Carol Sousa, and the Esri Press team.

Finally, we thank Tim Lewis, Green Infrastructure Center's editor, who provided text editing and endless patience.

About the authors

Karen E. Firehock

Karen E. Firehock is an environmental planner and natural resources manager who has worked on a diverse range of environmental issues for more than 30 years. In 2006, she cofounded the nonprofit Green Infrastructure Center Inc. to provide tools and methods for communities, agencies, and organizations to conserve and restore their natural assets. She is the Green Infrastructure Center's Executive Director and has worked on green infrastructure (GI) projects in many states at varying scales, ranging from state models to plans for multicounty regions, cities, small towns, and individual sites. She has also designed and installed habitat restoration projects in both rural and urban landscapes. For example, she designed an arboretum at the Hunter Holmes McGuire VA Medical Center in Richmond, Virginia, where she employed disabled veterans to design, plant, and manage the site to create a healing landscape.

Firehock has also been an educator for many years and teaches graduate courses in Green Infrastructure Planning at the University of Virginia. She also leads webinars for state, national, and international audiences. Her book *Strategic Green Infrastructure Planning: A Multi-Scale Approach* was published in 2015.

Firehock is also a leading practitioner in collaborative planning and has managed multiple community planning and engagement efforts for diverse topics such as landscape conservation, park master plans, heritage tourism, watershed plans, comprehensive plans, historic preservation, and sustainable development. She is also a trained mediator and has written books and manuals on community engagement, including *Collaboration: An Environmental Advocates Guide and Community-Based Collaboration*. She has also worked internationally, serving as the lead facilitator for a Middle East environmental collaboration workshop held in Jordan and the planning lead for University of Virginia students working on a water and health project in the Venda region of South Africa for 8 years.

Prior to directing the Green Infrastructure Center, she was a senior associate at the UVA Institute for Environmental Negotiation, where she managed community planning coalitions and wrote guides about stream buffer management, watershed planning, and water quality protection. For 12 years, she was the national Save Our Streams Program Director at the Izaak Walton League of America, where she directed stream and wetland conservation programs across the US and taught workshops in some 40 states, created the Stream Doctor™ habitat restoration program, and a wetlands-monitoring program and handbook. She also developed national volunteer biological monitoring protocols. She holds a Bachelor of Science in Natural Resources Management from the University of Maryland and a Master of Planning from the University of Virginia. She lives in the Piedmont region of Virginia in a historic 18th century house on the banks of the James River, and she enjoys kayaking down the rocky rivers of Virginia.

R. Andrew Walker

R. Andrew Walker is a GIS analyst, modeler, and land planner. He provided much of the graphics support and technical instructions for the mapping in this book. He has more than a decade of experience with high-level GIS analysis and modeling, with particular emphasis on using GIS to support urban and environmental planning applications.

He previously served as Green Infrastructure Center's senior GIS analyst, and currently works as a course developer at Esri. He has led dozens of spatial planning projects that support better decision making for conservation. Having also worked in developing nations for several years on a variety of spatial planning projects, he has facilitated diverse groups in creating strategic conservation and economic development plans. He has also taught advanced spatial analysis courses at the University of Virginia. Walker holds a Bachelor of Arts in Geography from the School of Geographical Sciences and Urban Planning at Arizona State University, and a Master of Urban and Environmental Planning from the University of Virginia.

Chapter 1

Green infrastructure: Considering nature before and during development

Decision makers today face the common challenge of determining the effects of development on the natural landscape within the context of profound and rapid change. Land development is a daily occurrence, but it often results in the loss of natural, life-sustaining resources that has consequences—not just for wildlife, but for drinking water, recreation, quality of life, and health. Even public safety is affected by the floods and fires that follow when land is disturbed or paved. Scientific evidence points to climate change, resulting in more severe weather events and rising sea levels that shift shorelines inland, causing habitat loss when artificial structures or paved surfaces create impediments to that movement. Sea level rise requires advance planning to accommodate these changes. When a landscape is denuded of trees and its natural waterways are impeded, changes to hydrology, air quality, fisheries, and water supply result. A clear example occurred in southeastern Texas in 2017, when Hurricane Harvey caused about $160 billion in damage, becoming the most expensive natural disaster in US history to date, according to AccuWeather.[1] The storm's severity resulted partly from increased temperatures in the Gulf of Mexico that caused the system to absorb record volumes of water. The water had nowhere to go, partly because of massive alterations to the landscape from paving and rechanneling of streams and marshes.

If communities want to become more resilient and learn to adapt to a changing climate, then we must discover—as Ian McHarg said back in 1969—how to *design with nature*.[2] This book shows how to design with nature by mapping, conserving, and restoring our most highly valued landscapes. This work will allow communities to become more resilient and understand the tremendous ecosystem services that our natural world provides.

Thanks to new analytical tools, we can measure, evaluate, and map our natural world more rapidly and widely than ever before. By identifying and integrating a network of critical landscapes, communities can protect the places and resources that help people, wildlife, and economies thrive. A key avenue to push that change in perspective is to begin to see our natural world as our *green infrastructure (GI)*.

Green infrastructure comprises the natural assets we see around us: trees, parks, streams, lakes, forests, and rivers. They form a natural system of interconnected ecological processes that protect our native species and provide human beings with many things essential to life: clean water, clean air, healthy lifestyles, and significant economic benefits. GI planning is not just about conserving our wildlife; it also seriously affects a community's social and economic health in ways that will be discussed throughout this book.

> ### What is green infrastructure?
>
> The science of green infrastructure provides a framework for sustainable growth and conservation, including such factors as:
>
> - **Protection of the Environment:** Green infrastructure protects the health and diversity of wildlife and maintains natural systems that deliver critical, life-sustaining services.
> - **Contributing to a thriving economy:** Green infrastructure benefits property values, lowers health care costs, boosts tourism, and helps communities make smarter investments in grey infrastructure.
> - **An enhanced quality of life:** Green infrastructure ensures people can connect with nature; have access to clean air and water; and live healthier, happier lives.

As a science, GI planning uses maps and other analysis and legal tools to plan for land development and conservation consistent with natural environmental patterns and the needs of developmental change. As such, GI planning becomes an invaluable way to promote both smart growth and smart conservation. Many communities already employ GI planning to improve their economies by using their natural assets to attract tourists, recreationists, and businesses to a healthful, attractive, and functional environment.

GI planning is also about protecting wildlands and wild places, even if no one ever visits them. Consider the vast populations of the western US that receive much of their water for drinking and agriculture from mountains and glaciers many hundreds of miles upslope. Wild places also need protection and consideration, even if they have no evident utility in our daily lives. Much as Aldo Leopold — the father of wildlife management in the US — promoted the land ethic in the 20th century, we encourage consideration for the natural values that our landscapes give us daily, both seen and unseen, known and unknown, in every planning decision.

The GI approach envisions decision makers and the public collaborating to preserve and link open spaces, watersheds, wildlife, habitats, parks, and other natural areas that enrich and sustain a community's quality of life, economy, and sense of place. This is a change from how planning has been practiced traditionally in the 20th century in the US and across the globe. Beginning planning with an assessment of natural resources and ecosystems as a first step rather than trying to provide open space and mitigate development after development plans have been put in place has resulted in the dysfunctional ecologies seen today. The flooding in Houston, noted earlier, and the tremendous fires in California in 2018 are a direct result of both climate change *and* poor planning that have not accounted for landscape connectivity and functionality. GI planning is a strategic approach that reorders traditional planning approaches by prioritizing the environment that sustains and enriches our lives first before development plans are made. Geographic information systems (GIS) users offer critical support for this work by mapping those critical environmental resources, such as groundwater recharge areas or fire-risk or flood-prone areas, and prioritizing their protection or avoidance. This book combines the skills and knowledge needed by GIS users with the tools and approaches of GI modeling and mapping to support the growing movement to conserve our world's GI.

Conservation means development as much as it does protection. I recognize the right and duty of this generation to develop and use the natural resources of our land; but I do not recognize the right to waste them, or to rob, by wasteful use, the generations that come after us.

Theodore Roosevelt, speech at Osawatomie, Kansas, August 31, 1910[3]

Canyon De Chelly National Monument, Arizona, protects both the natural and cultural landscape. Credit: Green Infrastructure Center Inc.

Why we need a new way to plan

Despite growing awareness of the significance of preserving our natural environment and a proliferation of new tools, planning with natural features and functions in mind is not yet widespread. We still develop in ways that disrupt natural systems. Sprawl-patterned development continues to consume natural landscapes, disconnecting wildlife corridors, causing excessive storm runoff and water pollution, marring scenic vistas, impacting historic and archeological sites, and paving over once productive aquifers, preventing their vital recharge.

> *Ecologists from across the globe estimated that humans have already transformed about 43 percent of the ice-free land surface of the planet, leaving the world's ecology at risk of collapse.*[4]

We need houses but often "grow" them, instead of food, on our best agricultural soils. We need clean water but often cover the tops of recharge zones with roads and parking lots. We must change our thinking to see our natural elements as part of our infrastructure and plan for development in ways that minimize land disturbance and maximize natural resource conservation. When we begin to think of nature as our GI because it provides us with our clean water and air, food and healthy lifestyle, and shaded homes and scenic vistas, we can intentionally develop infrastructure to foster conservation and safeguard natural elements that sustain and benefit us. In short, if we want healthy, economically vibrant communities, then we need healthy, vibrant landscapes too.

Species at risk

Despite many conservation victories, across America the natural landscapes and species that depend on them are increasingly isolated and put at risk. Endangered Species International estimates the number of endangered species for North America at more than 1,261, with many more species threatened or at risk.[5] As habitats become increasingly fragmented by roads, subdivisions, mines, and pipelines, biodiversity is put at risk when animals cannot reach new areas to take advantage of water or food or to find mates.

First, we must harness all the tools at our disposal to evaluate, map, and plan for our connected natural landscapes and resilient ecosystems within the context of sensibly planned growth and development. Fortunately, we now have the data and analytical tools to map large landscapes more accurately than ever before. Geographic information systems can integrate multiple datasets to craft a more sophisticated, detailed, and nuanced understanding of the natural landscape.

This book describes these new mapping tools and methods for GI planning that utilize the vast amounts of data now available. GIS analysts can access these data to organize and prioritize that information effectively and rapidly, saving time and money. These methods involve sophisticated modeling that turns scalable data into highly useful information directed to specific planning goals and projects. This book describes how to apply a GIS model for natural resource conservation, laying out the processes involved and explaining how these tools allow planners and decision makers to protect and foster a connected, resilient, and biologically diverse landscape in which both wildlife and humans can prosper. However, before beginning this journey of landscape conservation modeling, some context is needed.

Although still a relatively young country, the US has protected a great deal of land from development. With more than 400 national park areas, 560 national wildlife refuges, and nearly 250 million acres of other public lands managed by the US Department of the Interior, Americans have a wealth of publicly held resources supporting their natural assets and cultural treasures. And that doesn't include the thousands of state and local parks. Yet, because natural habitats don't end at park boundaries, both wildlife and humans require connections across the *entire* landscape. The growing demands of development can cause conflicts at these boundaries.

Using maps

Maps are a vital tool used extensively in GI planning. Maps enable planners to identify areas of natural resources and tag them with various criteria, as well as attach values to those criteria. The extent of specific natural resources, the balance of overall environmental assets, and their relationship to a range of overlays, such as leisure facilities, endangered species, future developments, the road and rail network, water recharge and supply sources, enable planners to identify both opportunities and potential problems.

Maps also are invaluable because they get the public involved. Highly visual maps can be printed and used in displays to identify public concerns, evaluate priorities, and help decision makers visualize issues at multiple scales. A local landscape can be seen within wider county, state, regional, national, and even international contexts. Maps can identify wildlife migration routes and nationally or regionally endangered species, show how roads fit within a regional network, and mark the location of a business park relative to a wider economic network.

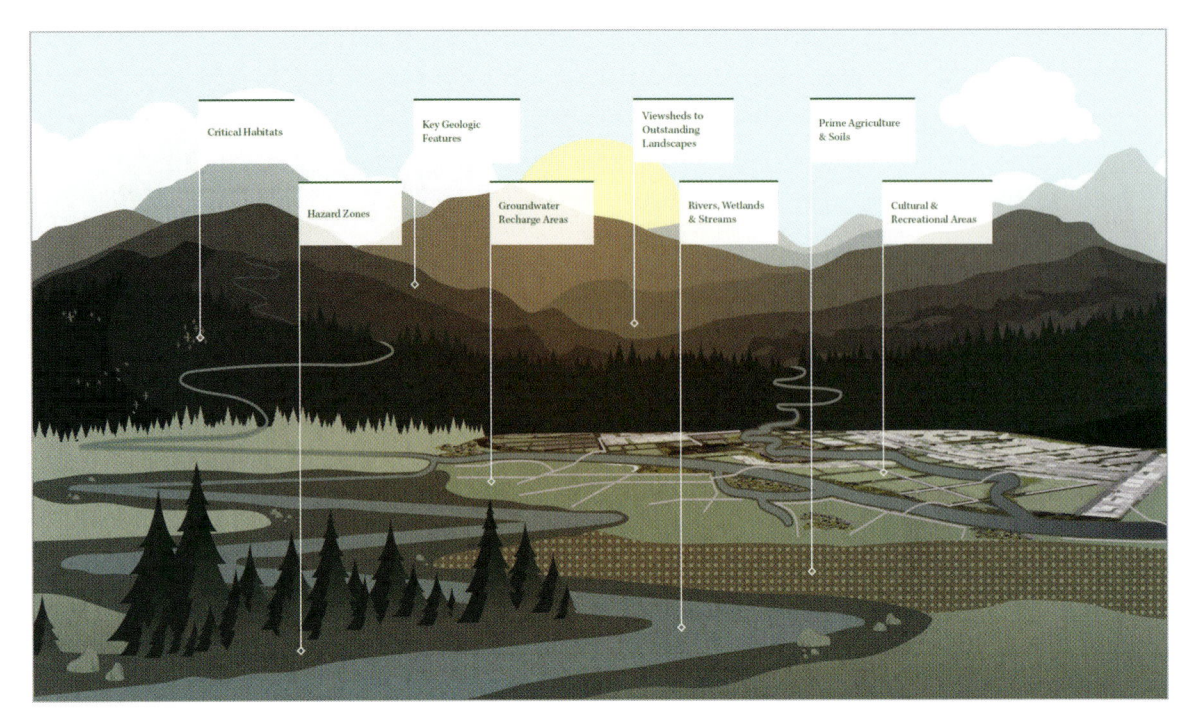

GI maps help communities grow smarter. They identify natural assets that allow communities to grow richer, healthier, and more responsibly while preserving valuable natural areas if they consider and identify the valuable green landscapes they wish to protect and connect. Credit: Esri.

A healthy landscape means a healthy community

People depend on a healthy landscape to sustain their very existence. Thoughtful, long-term planning intended to sustain the ecological services that nature provides clearly enhances a healthful community. Not planning for nature's benefits can have consequences. For example, a county in Virginia slated future housing developments to occur on its best agricultural soils, while designating areas with poor soils as the agricultural zone. Doing so put its food security at risk and foreclosed on its agricultural future. Why did this happen? Local planners did not use soil data to create maps that would have informed their decisions about where to locate future developments or agricultural uses. These problems are preventable. A healthy food supply for people and abundant habitat for wildlife can be maintained, side by side. But it requires good data and maps to inform planning decisions.

Undeveloped landscapes are not just important for food. People also need to access nature to remain healthy. Indeed, many studies show the importance of green spaces for both *mental* and *physical* health. Just being able to see green spaces can reduce illness and stress. In fact, one study found that employees without views of green spaces reported 23 percent more incidences of illness.[6] Hospital patients studied by Dr. Roger S. Ulrich at the Center for Health Systems & Design at Texas A&M University found that having views of nature led to faster recovery for patients. Patients experiencing views of nature had "shorter post-operative stays, fewer negative comments from nurses, took less pain medication and experienced mostly minor post-operative complications."[7] Many hospitals are beginning to provide rooms facing on green scenery or having photographs of nature on the walls; some hospitals are adding trails around their grounds. Most urban hospitals do not own the landscapes providing those views, so they depend on local planners and developers to maintain the green spaces that help their patients heal faster.

A vibrant tourist industry requires a vibrant natural landscape

Heritage tourism is another way that nature pays us back. Heritage tourists spend, on average, about 2.5 times more than other types of tourists. And these types of tourists often desire access to both culture and nature. In Nelson and Albemarle counties in Virginia, the Brew Ridge Trail follows the ridges and valleys that cross the piedmont of the Blue Ridge Mountains. The trail offers hiking and local craft beers within a natural landscape that is increasingly seen as the key factor in a strongly reviving tourist industry that also boasts wineries, bed and breakfast establishments, and destination weddings.

Similarly, in Colorado, the Brew Trail offers an interactive map that allows visitors to locate their brews and views and plan outings as part of an integrated leisure activity. These beer and nature lovers spend money on outdoor gear, local arts and crafts, hotel rooms, restaurants, and gas, as well as enjoying scenery and participating in outdoor and cultural activities.

Clean water requires a clean watershed

Beyond natural beauty, clean water is perhaps the most recognizable human value associated with landscape conservation. Protecting just 20 percent more forested land can reduce drinking water treatment costs by 10 percent, because water arrives cleaner to reservoirs or river intakes.[8] However, protecting clean water often requires interjurisdictional cooperation, because watersheds tend not to observe political boundaries. Rivers usually wander across multiple counties and through many urban areas, passing sewage treatment plants, factories, and storm water outlets. For example, Ulster County, New York, contains the upper Adirondack watersheds that supply the drinking water demanded by New York City's five boroughs. Healthy water that did not have to utilize expensive filtration systems because it is so clean is made possible by Ulster County's abundant forests, which have been protected for many years.

The focus of this book—GI planning—arises from a multitude of researchers who have studied how landscapes function and why connectivity is critical for creating resilient communities. For readers less familiar with how these theories arose, Appendix A reviews a brief history of the seminal work that led to the current status of the field and to the creation of the national GI model for the US hosted by Esri.

Key terminology in GI network design

Different disciplines use distinct and unique terms to refer to intact habitats. In landscape architecture, the term *patches* is often used to refer to a distinct habitat type. Scottish botanist and plant ecologist Alex Watt presented his theory of *patch dynamics* in his 1947 address to the British Ecological Society in which he suggested that plant species within bounded communities are distributed in *patches* that form mosaics across a landscape.[9] Patches can be described as distinct areas of similar types of habitat that differ significantly from adjacent landscapes and are often dependent on delicate climate factors or a specific underlying geology and hydrology.

Ecosystems can be thought of as a mosaic of patches. A grassland or mountain range is obviously a mosaic of many patches. A relatively intact and local forestland can also be considered a complex mosaic of patches, because it may also contain open, nonforested areas and wetlands. Patch size, shape, duration, and boundaries are mutable and of varying sizes and sensitivities to change. Disturbances such as wind, floods, and fire can irrevocably disrupt their communities. Although these disruptions can create openings in the habitat for new species, the distribution of the original species that utilized the disturbed area can be irreparably changed. For example, a blowdown in a forested area may become a meadow and provide forage areas for edge species, such as butterflies and birds, while displacing other species that reply on dense forest cover.

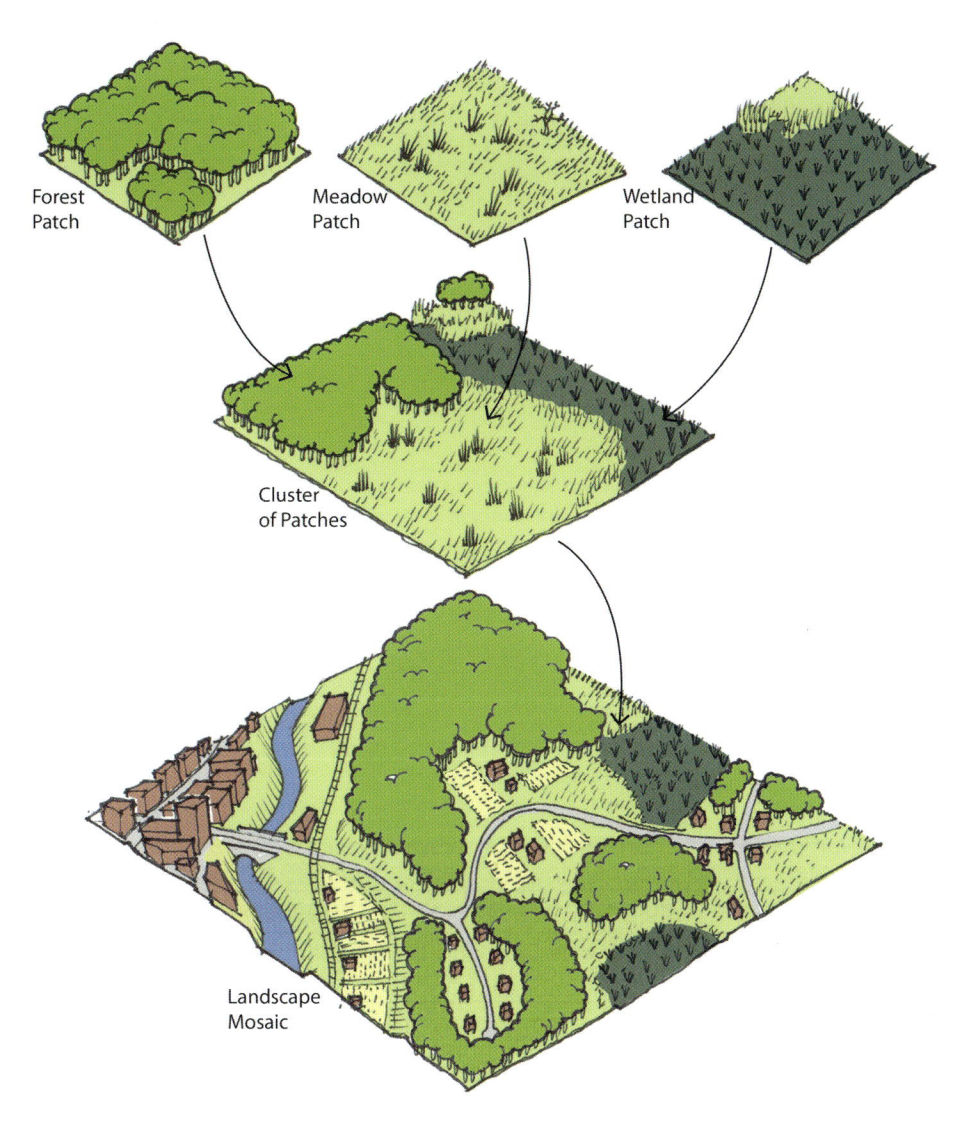

Mosaic is a term used to describe the pattern of patches, corridors, and boundaries with a matrix that forms an entire landscape. Credit: Green Infrastructure Center Inc.

Chapter 1 | Green infrastructure: Considering nature before and during development

Habitat fragmentation

Habitat fragmentation is the breaking up of natural landscapes into smaller and more disconnected pieces. Fragmentation leads to species loss and decline when damages to one area cannot be overcome by repopulation as organisms migrate to a new habitat. Similarly, areas where species have been lost are less likely to be recolonized when they become greatly isolated.

Conservation of the sage grouse

Although most land in the US is privately owned, concerted conservation efforts that involve multiple landowners across large landscapes can make a significant difference to land preservation, especially when it comes to species with a large but generally endangered range. For example, in the US, the greater sage grouse has recently been removed from its protected listing under the Endangered Species Act (ESA). According to former US Secretary of the Interior Sally Jewell, the determination that the greater sage grouse does not require ESA protection is "proof that we can conserve sage grouse habitat across the West while we encourage sustainable economic development."[10]

Thanks to coordinated restoration efforts, which have overcome fragmentation of habitat, sage grouse populations are now rebounding in the US. Image courtesy of Alan D. Wilson.

Habitat corridors

Central to the health and populations of animals and plants is the degree of connectivity between patches. Connectivity of natural habitats supports species movement by providing cover and forage area, as well as opportunities to reproduce and increase genetic diversity, thus contributing to species' viability over the long term. Corridors—vegetated linear areas of similar habitat types that differ from adjacent landscapes—allow sheltered passage from one patch to another, as well as forage and habitat. How well species can utilize corridors for movement depends in part on the individual species and on the habitat quality of the corridor itself.

Note: Corridors, and other technical terms of GI planning, are dealt with in more detail in chapter 2, "Modeling the natural landscape."

A corridor is primarily a linear pathway that connects habitat patches and is wide enough to allow for species to travel safely. Credit: Green Infrastructure Center Inc.

Planners often use the term *core* to refer to an intact landscape that is large enough to support a multitude of native species. The terms *core* and *patch* are often used synonymously. Cores are connected by corridors that provide pathways for species to move across the landscape.

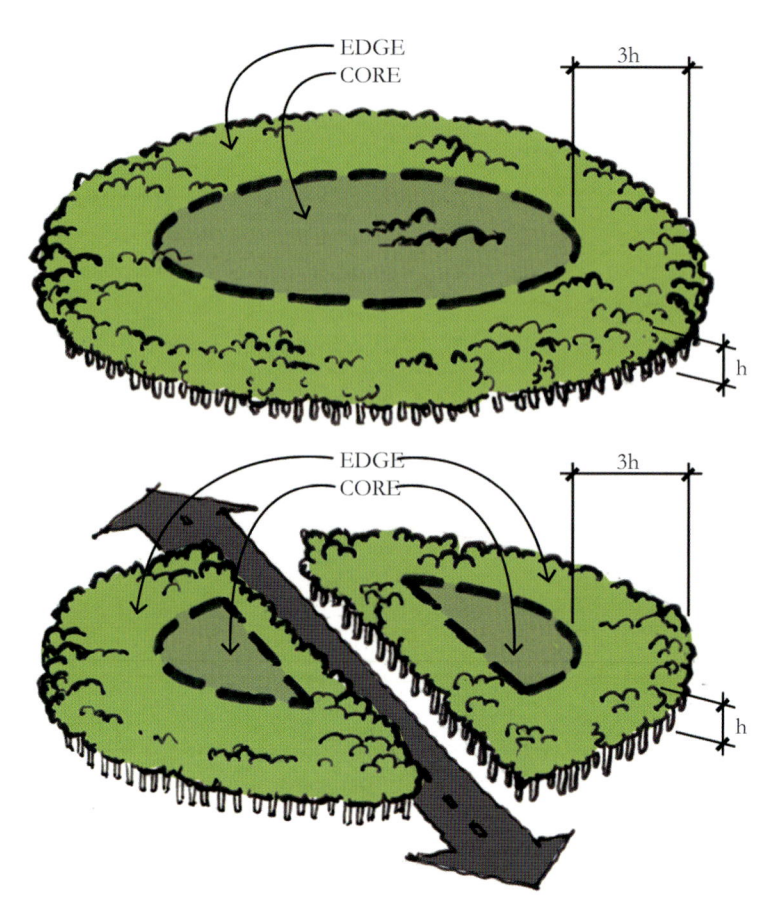

Edge area = Average tree height (h) X 3
Core = Total area - Edge area
Ideally, Core ≥ 100 acres

Habitat cores are measured by subtracting the edge and then determining if enough interior habitat remains to support a multitude of species.
Credit: Green Infrastructure Center Inc.

Cores (patches) and corridors have become key principles in the framework of connected landscapes. The term *core* comes from the work of the Biosphere Conference, which dealt with the utilization and preservation of genetic resources. In 1970, the United Nations Educational, Scientific and Cultural Organization (UNESCO) General Conference proposed a worldwide network of biosphere reserves to ensure genetic material is preserved across the globe. The conference also launched the Man and Biosphere (MAB) Program.[11] Each participating country had to designate specific biosphere reserves that would conserve key species and provide areas for research. They included criteria and guidelines for these reserves, particularly the importance of having core areas protected by buffer zones.[12]

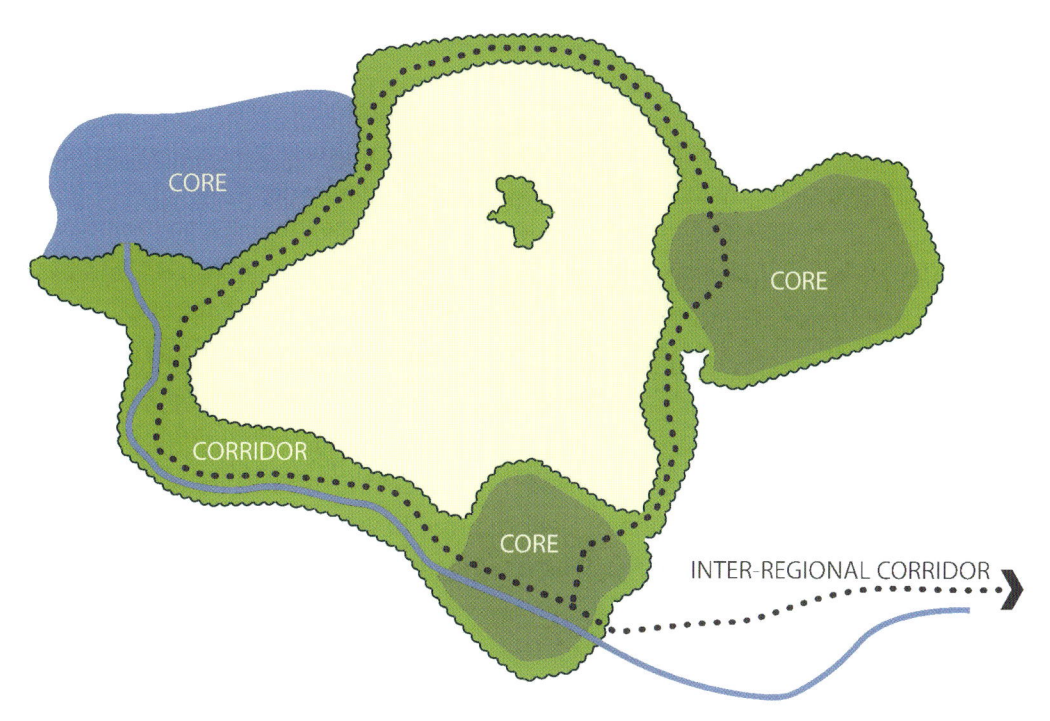

A GI network is composed of habitat cores and connecting corridors that support the biodiversity of both flora and fauna. Credit: Green Infrastructure Center Inc.

The UNESCO conference decided that a core needed to be large enough to meet the habitat needs of any species of concern in situ, whereas the buffer should be large enough to support some uses and provide space for research stations. In the design of biosphere reserves, cores (interior areas) are "securely protected sites for conserving biodiversity."[13] Cores can be considered as similar to patches, but they are also large enough to support more than one species and are surrounded by buffer zones and linked by corridors.

A GI network is an interconnected landscape of prioritized cores and connecting corridors.

Examples of biospheres in the US are the Everglades and Dry Tortugas, both in Florida, and the northern Mojave Desert, which encompasses four different management units of government land, including Death Valley National Park, Joshua Tree National Park, Anza-Borrego Desert State Park, and the Santa Rosa and San Jacinto Mountains National Monument.

Managing key biosphere ecosystems across multiple land ownerships presents challenges. For example, for a long time, the Dana Biosphere Reserve in Jordan faced challenges because of poaching, as its previous uses included subsistence hunting by local people over many centuries. The problem was solved only when local hunters were trained as nature guides, and today they protect the wildlife rather than eat it. Other joint management areas, such as the Changbai Mountains bordering China and North Korea, face challenges related to managing species across borders of countries that lack relationships stable enough to facilitate shared conservation planning.

The Dana Biosphere Reserve in the Hashemite Kingdom of Jordan supports many rare and endangered species within 320 square kilometers along the face of the Great Rift Valley. It encompasses four different biogeographical zones: Mediterranean, Irano-Turanian, Saharo-Arabian, and Sudanian. Credit: Green Infrastructure Center Inc.

Corridors provide a way for wildlife to cross the landscape and increase potential connectivity between habitat patches, which allows for greater movement and intermingling of populations. This intermingling, in turn, allows for greater genetic diversity, as well as for repopulation of previously disturbed areas. The role of corridors for providing connectivity has increasingly been recognized, as has the understanding that the corridors themselves provide habitat.

GI modeling in the US

The very idea to call systems of habitat cores and connecting corridors "green infrastructure" can trace its origins to the Florida Greenways Commission. The focus on connecting green resources followed work focusing on important natural areas in Florida led by The Nature Conservancy, as well as advocacy and planning by the Conservation Fund and 1000 Friends of Florida. The commission studied the Loxahatchee River and nearby conservation lands in southeast Florida as a connected landscape that represented some of the most pristine habitat remaining in that rapidly developing part of Florida. Based on this work, in 1991, 1000 Friends of Florida and their partners advocated for a statewide green network covering the entire state. As a result, Governor Lawton Chiles appointed the Greenways Commission for that purpose. Its 1994 report proposed a linked habitat network, using Reed F. Noss's recommended statewide design as a foundation for a network of core preserves, buffer zones, and corridors.[14]

The report also laid out the Greenway Commission's vision of the state's natural resources as infrastructure: "The Commission's vision for Florida represents a new way of looking at conservation, an approach that emphasizes the interconnectedness of both our natural systems and our common goals and recognizes that the state's 'GI' is just as important to conserve and manage as our built infrastructure."[15] A team at the University of Florida created a network of cores and corridors resulting in the Florida Ecological Network design.[16] For more on this model, see chapter 2.

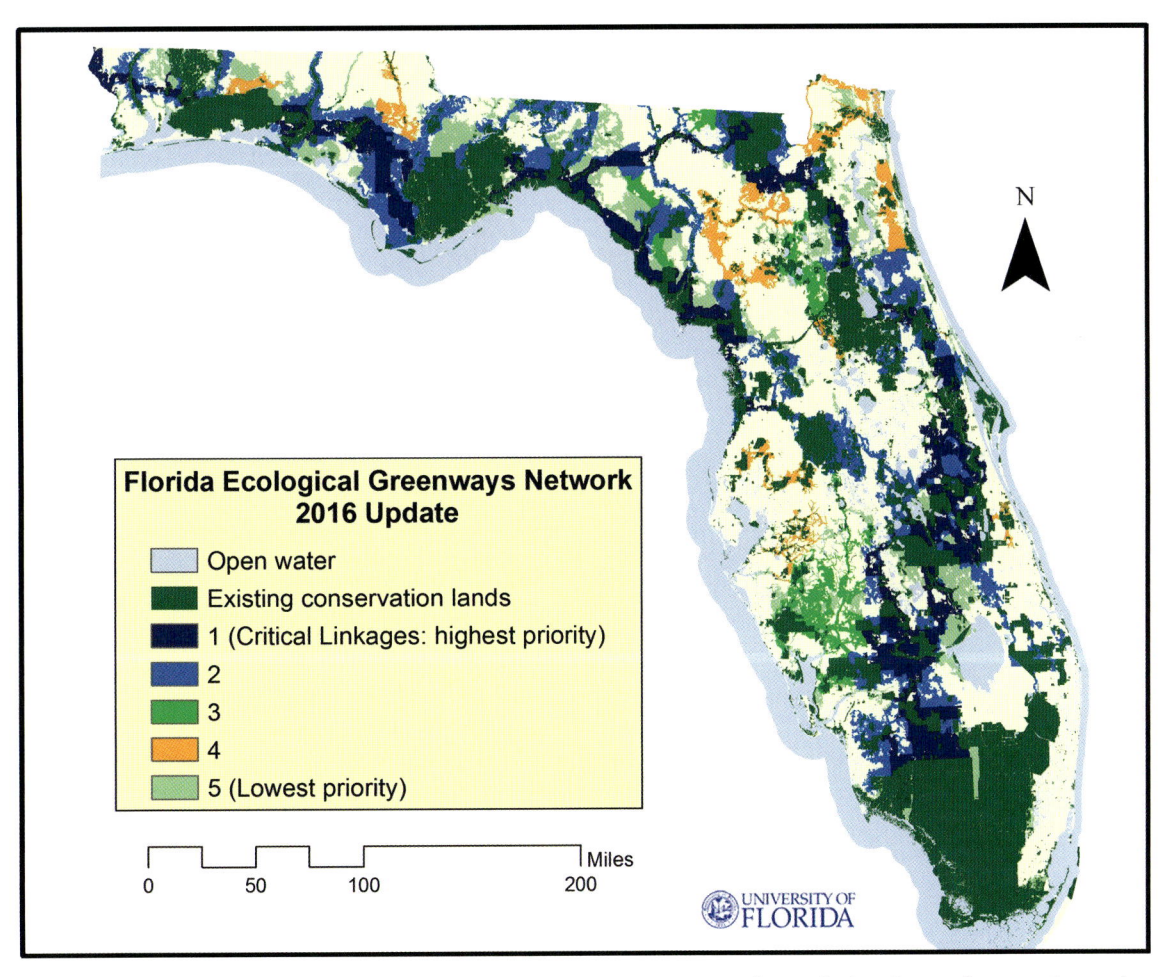

Florida Ecological Greenways Network. Credit: Tom Hoctor, Ph.D., Director, Center for Landscape Conservation and Planning, University of Florida.

Similar to Florida, Maryland's GreenPrint Program originated with an emphasis on greenways. The Maryland Greenways Commission was established in 1991; its purpose was to create a statewide network of greenways that would provide natural pathways for wildlife movement and trails for recreation and alternative transportation routes. The assessment was based on Florida's ecological network approach.

In 1997, the Maryland Department of Natural Resources provided Baltimore County with a grant to develop a county model of GI. In 1999, during his inaugural address, Maryland governor Parris Glendening called for the state to plan carefully for its forests, woodlands, streams, and rivers as integral parts of Maryland's GI. He explained that this GI was just as important as the state's roads and bridges.[17]

In 2000, the Maryland Department of Natural Resources used the prototype developed for Baltimore to create a state GI network of hubs and corridors, which today is called the Maryland GreenPrint, of which only 25 percent of those lands are protected. To better guide conservation of these lands, the GreenPrint map depicts Targeted Ecological Areas (TEAs), lands and watersheds of high ecological value identified as conservation priorities by the Maryland Department of Natural Resources. It also displays priorities for the state's land conservation programs: Program Open Space, the Maryland Agricultural Land Preservation Foundation, the Maryland Environmental Trust, and the Rural Legacy Program. GreenPrint helps these programs integrate their priorities and helps to steer land acquisitions for Program Open Space. The TEAs were first developed in 2008 and then updated in 2011.

Since the development of the Florida and Maryland models, statewide GIS-based GI models have been built by state agencies for Virginia, California, Colorado, and Montana. The Green Infrastructure Center built statewide models for New York, Arkansas, and South Carolina. The Conservation Fund built a statewide model for West Virginia and has also built many regional models. In 2006, the Conservation Fund further popularized the concept in its book *Green Infrastructure*, which describes GI as "a strategically planned and managed network of wilderness, parks, greenways, conservation easements, and working lands."[18] Other groups, such as The Trust for Public Land, The Wilderness Society, Defenders of Wildlife, and The Nature Conservancy (which founded a conservation modeling firm, NatureServe), as well as many universities and local groups, are conducting GI planning while developing new models, tools, and methods to do this work.

Most statewide models, and the national Esri model, were constructed using land cover at a 30-meter resolution. However, local governments have also created higher resolution models at the 1-meter scale to allow for more detailed assessments at the local level.

The risk of isolating populations of species

Isolated populations are most at risk of depredation or extinction. If a disease affects a species, causing it to decline or disappear, isolated areas are less likely to be recolonized. Also, such smaller habitat areas cannot support abundant species.

For example, although American brown bears (*Ursus arctos*), also commonly referred to as grizzly bears, are abundant in Yellowstone National Park, they occupy less than 2 percent of their former range. Bighorn sheep (*Ovis canadensis*) are endangered and have declined from an estimated population of 1.5 million to 2 million to only 70,000 today. Some species, far less noticeable, such as the Karner blue butterfly (*Lycaeides melissa samuelis*), are also declining. Unlike large megafauna, small insects may not capture our attention as readily, and yet they form a vital part of the ecosystem as pollinators, thereby indirectly contributing to the beauty of the world's landscapes and feeding the world.

Bighorn sheep (*Ovis canadensis*) populations, although eliminated from Washington, Oregon, Texas, North Dakota, South Dakota, Nebraska, and part of Mexico in the 1920s, are now rebounding in areas where they had been reintroduced. Credit: Image courtesy Alan D. Wilson.

In the US, sprawl—wasteful patterns of land development that place people in areas farther from built areas and cities—creates landscapes that require more road building, causing loss of wildlife habitat and reducing agricultural lands. Part of this destruction is driven by a lack of consideration for the natural resources that are being lost.

In late 2015, the Green Infrastructure Center partnered with Esri to create a national model of cores using the script Green Infrastructure Center had created for its South Carolina GI model. Esri has since developed online apps to help users visualize different priorities for GI networks. As part of this process, Esri convened a technical committee made up of representatives from Green Infrastructure Center, NatureServe, the Conservation Fund, The Trust for Public Land, and Conservation Science Partners, all of whom reviewed the model. In several cases, surrogate data were needed because not all the data that a state model might supply were available nationally. The types of data used to create the national model are covered in chapter 4, whereas chapter 5 includes tips for creating customized GI models at the user's desired scale.

The national GI model for the US provides, for the first time, a countrywide model that can be used for conservation planning at the county, region, state, and national levels. However, to refine the model for local use, users should add more data. The national GI model is available for download at the state scale; data can then be clipped to the user's area of interest, such as a county, region, or watershed. The online apps that come with the model are especially useful to help users test quick snapshots of an area or for goal setting. Chapter 2 provides the description of the model and its component parts.

Green infrastructure planning around the world

Key foundational principles for GI mapping and planning originated from different countries. The early work of visionaries and scientists such as Alexander von Humboldt and landscape designers such as Calvert Vaux played a significant role. See Appendix A for more on the foundational work that led to GI planning.

In some countries, GI plans arise entirely from the grassroots, led by nongovernmental organizations. In 2010, the Tenth Meeting of the Conference of the Parties to the Convention on Biological Diversity held in Japan set the stage for the Strategic Plan for Biodiversity 2011–2020. In 2015, the Ecological Society of Japan, the largest membership-based conservation organization in the country, sent a delegation to the US to study GI planning. They hoped to address the increasing habitat fragmentation in their multi-island nation and protect the country's rich natural assets.

Ecological Society of Japan leaders and Green Infrastructure Center staff study the GI network along Virginia's Blue Ridge Mountains. Credit: Green Infrastructure Center Inc.

Meanwhile, efforts in Central America have focused on the conservation needs of particular species. One such project was the Jaguar Corridor Initiative. Partners, including the US-based Wildlife Conservation Society and various Central American groups, started the project in 1990 to link jaguar populations from northern Argentina to Mexico. In many cases, protecting the habitat of one animal that serves as an umbrella species also protects many more species. The Jaguar Corridor Initiative identified 44 corridors that were threatened by human uses, and one specific area in Costa Rica, the Barbilla-Destierro Biological Corridor, was determined to be particularly important to preserving the jaguar's genetic diversity.

In European Union (EU) countries, national governments are driving forces in motivating GI planning and public-private partnerships are forming to map and plan for GI across the EU.

In 1993, the EU created the Convention on Biological Diversity (CBD), with three main objectives:

- Conservation of biological diversity
- Sustainable use of the components of biological diversity
- Fair and equitable sharing of the benefits arising out of the utilization of genetic resources

In 2006, the European Centre for Nature Conservation published the *National Ecological Networks of European Countries Map*. As part of its Europe 2020 strategy, the EU created a plan to reverse biodiversity loss and move toward a resource-efficient, green economy.[19] More recently, the CBD developed the 20 Aichi Biodiversity Targets, of which Strategic Goal C is "to improve the status of biodiversity by safeguarding ecosystems, species, and genetic diversity," whereas Target 17 is especially relevant to GI, as it calls for each signatory country to have an effective, participatory, and updated national biodiversity strategy and action plan. The principles also recommend that planning take place across country boundaries to ensure the survival and health of endangered or at-risk species. The vision is that "by 2050, biodiversity is valued, conserved, restored and wisely used, maintaining ecosystem services, sustaining a healthy planet and delivering benefits essential for all people."[20]

The European Commission's (EC) biodiversity strategy recognizes the impacts of land fragmentation, finding that nearly 30 percent of the EU territory is moderately to highly fragmented. In Target 2 of its protocol, the EC recommends maintaining and enhancing ecosystem services and restoring degraded ecosystems by incorporating GI in spatial planning.[21]

A 2016 review of national ecosystem assessments compiled for the EU showed that many assessments were developed to meet the goals of the EC's biodiversity strategy. The assessments focused primarily on ecosystem services values, such as providing clean water, although some of the assessments also regarded biodiversity as a value in and of itself. Of those, Portugal, the United Kingdom, Spain, and Flanders conducted the most comprehensive studies. The United Kingdom, Flanders, Norway, and the Netherlands also addressed cross-border ecosystem services.[22] Some countries put greater emphasis on stakeholder engagement than others.

Researchers have recommended the EU focus on standardization of data collection, indicators, and methods to assess biodiversity and ecosystem services. This focus is especially important, given the many ecosystems that span multiple national boundaries, such as the Pyrenees between France and Spain; the Alps between Switzerland, Austria, France, and Italy; and the Carpathians between Hungary and Romania.

Several researchers have proposed a pan-European Green Infrastructure Network that would overlay quantified ecosystem services for specific landscapes with identified core habitats and wildlife corridors.[23] They mapped a large region of Europe for its capacity to regulate air quality, water flows, erosion protection, coastal protection, crop pollination, soil protection, water purification, and climate. These topical overlay maps could be combined to find areas satisfying the greatest number of ecosystem services. They also modeled essential cores using the Landsat Vegetation Continuous Fields tree cover layer from the Global Land Cover Facility to locate core habitats of at least 50 percent forest density and 500 square kilometers in size. Some countries showed extensive areas of core habitats for large mammals (approximately half of Estonia, Slovenia, Latvia, and Austria), whereas others, such as Cyprus and Denmark, had none.[24] The authors acknowledge that their academic endeavor did not engage governments in vetting the priorities but does show the potential that GI maps offer the EU.

Another driver for on-the-ground conservation work in Western and Central Europe is provided by the goals of the EC's Natura 2000 initiative. As part of this initiative, the EU committed to providing support and a legal framework for the preservation of national habitat networks. Natura 2000 is a network of core breeding and resting sites for rare and threatened species and for some rare natural habitat types. Stretching across all 28 EU countries, including land and aquatic resources, it aims to ensure the long-term survival of the region's most valuable and threatened species and habitats.[25] A more inclusive, multistakeholder effort has been implemented to meet Natura 2000's goals for a European greenbelt. This effort is discussed further in chapter 8.

Esri's national GI model creates a key tool for land planning in the US. By providing the analysis and location for large intact habitat cores across the country, the national GI model provides a critical starting point for geographers, planners, landscape ecologists, and architects to develop their own conservation plans. However, the Esri model is just the beginning. It is intended for GIS analysts to download the model, update or add data, change priorities, or uncover new relationships that inform conservation planning. Chapter 2 provides more information on the scientific underpinnings of the US national GI model; chapter 3 discusses the essential process for establishing goals. Chapters 4 through 8 then describe how to obtain the right data and utilize the model, create maps, identify risks, discover opportunities, and, most important, implement plans.

Notes

1. "Accuweather predicts Hurricane Harvey to be more costly than Katrina, Sandy combined," September 1, 2017. https://www.accuweather.com/en/weather-news/accuweather-predicts-hurricane-harvey-to-be-the-most-costly-natural-disaster-in-us-history/70002597.
2. McHarg, Ian L., and Lewis Mumford, *Design with Nature* (Garden City, NY: The Natural History Press for the American Museum of Natural History, 1969).
3. "Theodore Roosevelt." http://www.theodore-roosevelt.com/trenv.html. Website accessed February 2018. Visit the website to see the full range of Roosevelt's conservation ideas and effort.
4. Barnosky, Anthony D., Elizabeth A. Hadly, Jordi Bascompte, Eric L. Berlow, James H. Brown, Mikael Fortelius, Wayne M. Getz, et al., "Approaching a state shift in Earth's biosphere." *Nature* 486, no. 7401 (2012): 52–58.
5. Endangered Species International. http://www.endangeredspeciesinternational.org/overview2.html. Website accessed June 2017.
6. Kaplan, Rachel, and Stephen Kaplan, *The Experience of Nature: A Psychological Perspective* (Cambridge: Cambridge University Press, 1989).
7. Ulrich, Roger, "View through a window may influence recovery." *Science* 224, no. 4647 (1984): 224–225.
8. Ernst, C., *Protecting the Source: Land Conservation and the Future of America's Drinking Water* (Washington, DC: The Trust for Public Land, 2004).

9. Watt, Alex S., "Pattern and process in the plant community." *Journal of Ecology* 35, no. 1/2 (1947): 1–22.
10. Office of the Secretary, US Department of the Interior, "Successful conservation partnership keeps bi-state sage-grouse off endangered species list." Press release, April 21, 2015. **https://www.fws.gov/greatersagegrouse/Bi-State/04-21-15%20Bi-State%20Sage%20Grouse%20FINAL.pdf**. Website accessed September 2016.
11. Batisse, Michel, "The biosphere reserve: A tool for environmental conservation and management." *Environmental Conservation* 9, no. 2 (1982): 101–111.
12. *Task Force on Criteria and Guidelines for the Choice and Establishment of Biosphere Reserves. Final Report: Paris, 20–24 May 1974* (Paris: UNESCO, 1974).
13. Ibid.
14. Noss, Reed F., Michael O'Connell, and Dennis D. Murphy, *The Science of Conservation Planning: Habitat Conservation under the Endangered Species Act* (Washington, DC: Island Press, 1997).
15. Florida Greenways Commission, *Report to the Governor: Creating a Statewide Greenways System: For people . . . for Wildlife . . . for Florida* (Tallahassee, FL: Florida Department of Environmental Protection, 1994).
16. Hoctor, Thomas S., Margaret H. Carr, and Paul D. Zwick, "Identifying a linked reserve system using a regional landscape approach: The Florida ecological network." *Conservation Biology* 14, no. 4 (2000): 984–1000.
17. *Maryland's Green Infrastructure Assessment and GreenPrint Program* (Arlington, VA: The Conservation Fund. Case Study Series, 2004). **http://www.conservationfund.org/images/programs/files/Marylands_Green_Infrastructure_Assessment_and_Greenprint_Program.pdf**. Website accessed January 2016.
18. Benedict, Mark A., and Edward T. McMahon, *Green Infrastructure: Linking Landscapes and Communities* (Washington, DC: Island Press 2006).
19. European Commission, "Communication from the Commission to the European Parliament, the Council, the Economic and Social Committee and the Committee of the Regions, Our life insurance, our natural capital: An EU biodiversity strategy to 2020," (2011/2307[INI]), 3.
20. "Aichi biodiversity targets." **https://www.cbd.int/sp/targets/#GoalA**. Website accessed February 2018.
21. European Commission, "Communication from the Commission to the European Parliament."
22. Schröter, Matthias, Christian Albert, Alexandra Marques, Wolke Tobon, Sandra Lavorel, Joachim Maes, Claire Brown, Stefan Klotz, and Aletta Bonn, "National ecosystem assessments in Europe: A review." *BioScience* 66, no. 10 (2016): 813–828.
23. Liquete, Camino, Stefan Kleeschulte, Gorm Dige, Joachim Maes, Bruna Grizzetti, Branislav Olah, and Grazia Zulian, "Mapping green infrastructure based on ecosystem services and ecological networks: A pan-European case study." *Environmental Science & Policy* 54 (2015): 268–280.
24. Ibid.
25. Ibid.

Chapter 2

Modeling the natural landscape

This chapter presents a primer on the ecological concepts that inform the design of a green infrastructure (GI) network to help communities protect places and natural resources. It outlines the basic principles and terms of GI planning, especially as they relate to the various features of the landscape central to the GI planning process as well as the Esri National Green Infrastructure Model. To learn about the data collection process in greater detail, including the use of these methods to analyze the habitat cores data in a specific area of interest and how to model the natural landscape using geographic information systems (GIS)-based software, see chapter 4.

Esri launched the Green Infrastructure Initiative to assemble, evaluate, and produce resources that will enable governments and other entities to engage in GI planning while guiding development decisions. The Green Infrastructure Initiative is a collection of authoritative geospatial resources, newly generated data, online applications, and a downloadable model with the aim of empowering local organizations engaged in GI work, all while initiating a national vision of GI planning.

Esri created a habitat cores model for the US as part of its initiative to support GI planning. The model was created in collaboration with the Green Infrastructure Center and an advisory panel composed of experts in ecological modeling and conservation science. This new data tool provides a wealth of new information to those engaged in land-use planning, conservation, and natural resource management. Details on how to access the model are provided later in this chapter.

Why we need landscape models

GI benefits our lives and supports the landscapes where we live in many ways:

- **Protects the environment:** GI protects the health and diversity of wildlife and maintains natural systems that deliver critical, life-sustaining services.

- **Helps the economy thrive:** GI benefits property values, lowers health care costs, and helps communities make smarter investments in gray infrastructure.

- **Enhances quality of life:** GI ensures that people connect with nature, have access to clean air and water, and live healthier, happier lives.

Green infrastructure encompasses much more than river greenways or green corridors. Although GI planning utilizes vegetated *corridors* as critical connectors between habitats, it sets them within a wider ecological context. Rather than regarding the corridors as the focal point of a green strategy, GI emphasizes the role of those corridors as links between larger blocks of intact habitat that provide sizable, wildlife-sustaining habitat cores capable of supporting a diversity of species. It places a significant value on these core habitats, depending on their integrity, size, and quality. Corridors are important, but without the cores, there is significantly less overall diversity in the landscape. Planners sometimes prefer the terminology of *hubs*, *cores*, *links*, and *sites*, whereas landscape architects refer to *patches*, *corridors*, and *sites*. In this book, we use the common references of *cores*, *corridors*, and *sites*, but regardless of the terminology, the principle is still to conserve large blocks of intact habitat, connected by corridors that allow for the movement of species. Species use the corridors to forage, nest, breed, and move between core areas.[1–5]

The purpose of landscape-scale GI models is to provide a context for identifying and mapping existing and future natural resources using the best information available. Such models provide planners and other decision makers with more complete information on which to base their decisions—better information for better decisions.

GI models also provide a baseline inventory of landscape features worthy of preservation and management and an estimate of their quality. Having a baseline of current land cover to compare with future land cover allows for analysis of change over time: Consider whether we are optimizing and conserving our use of natural or agricultural lands. Are we using maps to see how natural assets are distributed, and whether connections can be created between them? Are we determining whether habitat has been lost or become more fragmented? Are we looking for spatial patterns for growth or conservation?

A GI basemap may not include all the information that would be desired in a perfect world (it can be costly to collect ecological data from large areas), but new data can and should be incorporated as they become available. Indeed, having a baseline map can highlight gaps in our knowledge and justify why new data are needed. For example, the GI model shows an area likely to provide an ideal habitat for amphibians, because it has extensive water features. The GI model can thus be useful for pinpointing where specific investigations could be needed in the future.

With a GI strategy, one can preserve and connect open spaces, watersheds, agricultural resources, wildlife habitats, parks, and other critical landscapes—while avoiding hazards. Landscapes are often complex. However, we can find useful methods for distilling the available information to show a landscape's major patterns and characteristics.

Building a GI model

This chapter continues by defining the basic terms used in a GI model, describing how they are used, and providing examples. It then delves into the process of collecting, collating, and storing data, giving detailed information on sources and methods, including sections of code. Finally, it shows how to build a model and adapt it to specific goals and outcomes.

GI landscape terminology

In this book, the term *landscape* is defined as a geographic expanse containing multiple land covers and associated ecosystems.

The term *landscape-scale* is commonly defined as a scale that is appropriate for generalizing the phenomena that occur across a defined area of interest, which could be a watershed, county, state, or entire country. It can be contrasted with *site-scale*, which is a scale appropriate for assessing individual sites.

Land cover literally is what covers the land, generally vegetation types, water, or hard surfaces. In GI maps, stress is put on habitat types covering the land, such as forests, wetlands, and dunes. In the US, most GI models utilize the National Land Cover Database (NLCD).

A *patch* is a relatively homogeneous, nonlinear area of natural cover that differs from its surroundings. It merely refers to different types of land cover and does not relate to its integrity.

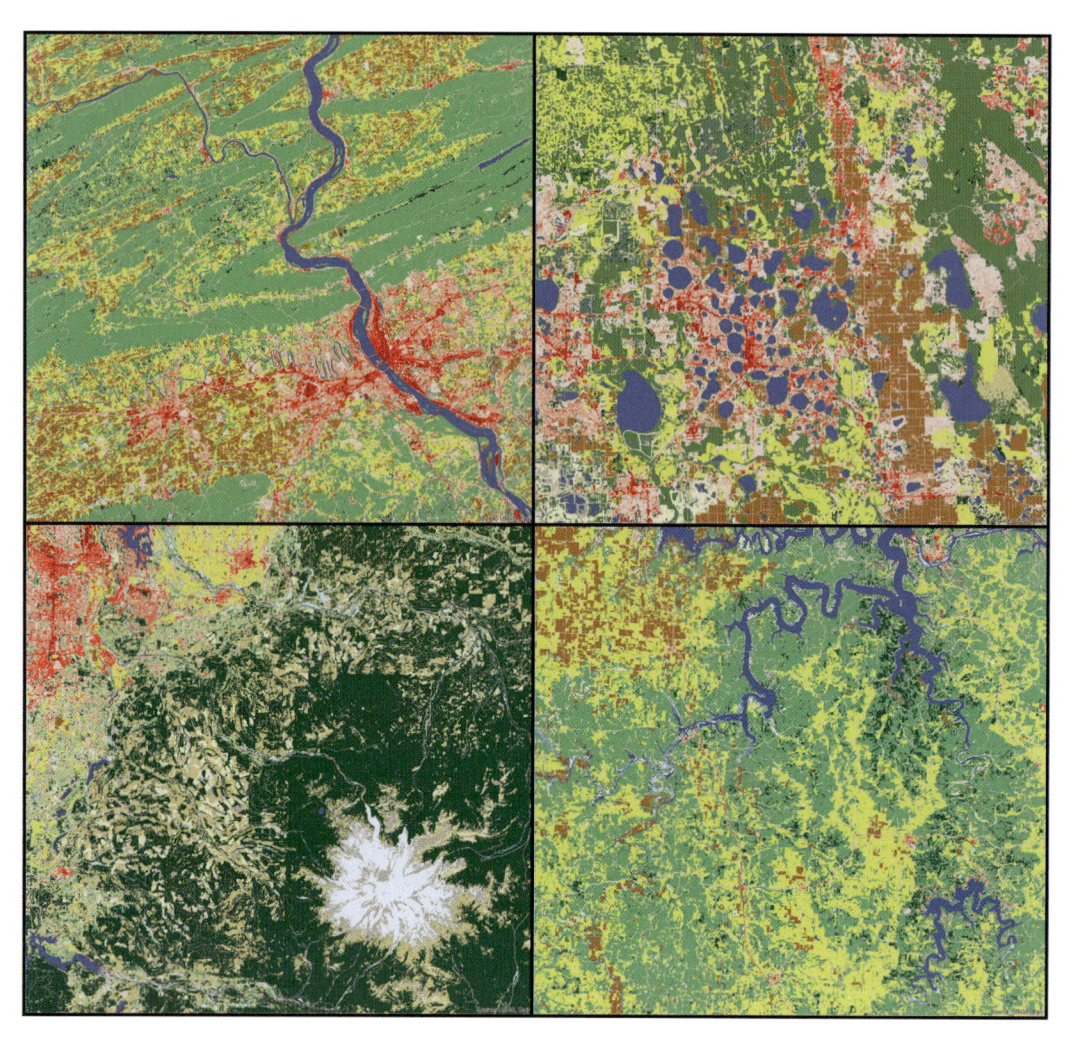

Land cover patterns and characteristics differ greatly and will influence resulting analyses. Credit: Green Infrastructure Center Inc. Data from the National Land Cover Database, US Geological Survey.

A *habitat* is the natural home or environment of an animal, plant, or other organism in which a species lives and makes up the locality in which it can find food, shelter, protection, and mates for reproduction. The physical factors of a habitat include soil type, underlying geology, hydrology, temperature range, and seasonal changes, as well as biotic factors such as the availability of food and the presence of predators. Every organism requires certain habitats for it to thrive. In practice, many species' habitats tend to be integrated within a single landscape, or *ecosystem*, providing a complex interconnectedness in which changes in uses or behavior by one species may affect others.

Intactness and fragmentation

Chapter 1 highlighted the issue of *habitat fragmentation*, which is a significant factor in species health because it affects the protection of the species and the minimization of risk from edge effects. *Ecological integrity* is the ability of an ecosystem to maintain those natural processes that sustain the species it supports. It is quantified according to many factors, most notably the size of the habitat, the extent of fragmentation, the depth to its interior, the area-to-perimeter ratio, soil diversity, elevation changes, water availability, and whether the area also supports rare or endangered species. Ecosystems with high integrity are large, have a deep interior, and are water rich. They function well as a habitat and thus provide more life-sustaining benefits to the species they support. These well-vegetated natural landscapes also provide other functions, such as holding water that can recharge aquifers or preventing erosion.

> *Ecological integrity is primarily determined by the size of an area of habitat, the extent of any fragmentation within it, and its edge-to-interior size ratio (the area-to-perimeter ratio).*

Landscapes become less intact as they become more fragmented. *Intactness* relates to the size and integrity of a habitat—the degree to which it is fragmented or not. Fragmentation can be directly measured and is assessed in terms of a ratio of intact habitat to overall landscape area. A national forest covering 100 square miles might have only 70 square miles of intact forest in several large cores, whereas the rest of the area is fragmented by roads, old remnant farms, and the like. The intactness of the landscape also depends on the existence of corridors between those cores and the ease with which species can move between them. The size of the habitat cores is also a primary factor in ranking their integrity.

Ecological integrity

Ecological integrity is the ability of an ecological system to support and maintain a community of organisms whose species composition, diversity, and functional organization are comparable to those of natural habitats within a region. An ecological system has integrity when its dominant ecological characteristics occur within their natural ranges of variation and can withstand and recover from most perturbations imposed by natural environmental dynamics or human disruptions.[6]

The *intactness* of natural lands is a key underpinning for the ecological integrity of a given area. An expanse with large, contiguous tracts of natural land will exhibit a high degree of intactness. Conversely, an area with small, isolated patches of natural lands will demonstrate a high degree of *fragmentation*. Fragmentation has an *exponential effect* on species survival.

Breaking up a 1,000-acre core habitat with a road that occupies only 20 acres of land surface will diminish a far larger acreage of habitat when edge impacts from the road are considered. Indeed, for some species highly sensitive to human intrusion, the road might destroy the habitat.

Another important concept is *isolation*. Fragmentation can lead to a local population of a species becoming isolated from other members of its species. A large road or housing development may isolate even a large natural area from repopulation from outside. Imagine the effect a dam has on upstream fish populations. A natural forest area might be quite large, but if new individuals of a species, such as wolves, cannot supplement an isolated population, that species becomes highly vulnerable to extinction. Plant and animal populations will naturally fluctuate, some to a high degree. However, species with smaller starting populations, especially those of indicator species, are more vulnerable to these fluctuations, and an isolated local population is more likely to become extinct.[7–9]

Human activities are a common source of habitat fragmentation. For example, urbanization, logging, mining, and road building are just a few of the ways that human activities fragment natural landscapes. As noise is a significant disturbance factor for many species, activities involving, for example, all-terrain vehicles and snow mobiles that create loud disturbances can affect species. Habitat fragmentation is one of the primary threats to wildlife and biodiversity, as well as being a key driver in species extinction.[10,11] Intactness and fragmentation are quantified using GIS, and specific programs such as FRAGSTATS 4 and Patch Analyst 5 have been developed to assess fragmentation.

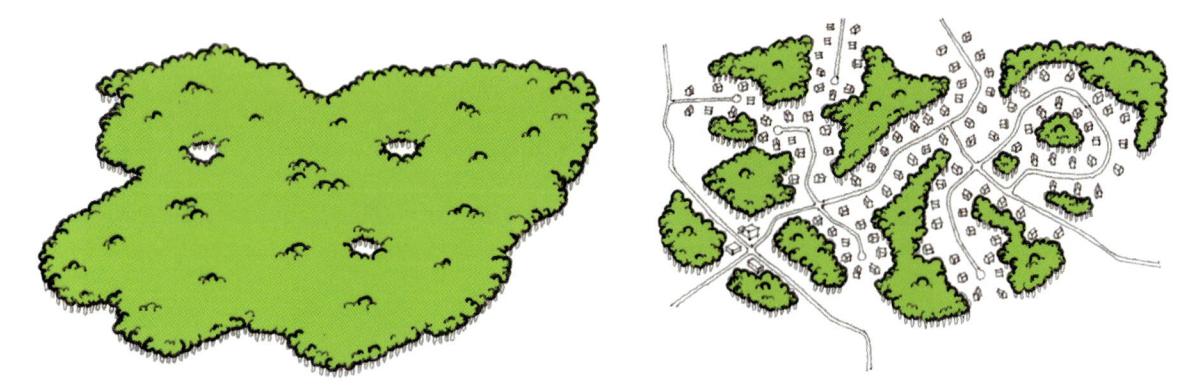

The first core (*left*) shows an intact core, and the second core (*right*) has become fragmented. Natural lands are often fragmented by the construction of roads and buildings, which creates smaller, more isolated patches of habitat. Credit: Green Infrastructure Center Inc.

A landscape is a *complex* of interlocking and overlapping habitats of individual species that are often mutually dependent and have some sort of symbiotic relationship. Thus, a landscape can become denuded of multiple species if one species becomes extinct—or, if the extinct species was a predator, another species can multiply out of control and become an extreme nuisance and even overcrowd and overgraze a landscape and cause further degradation. The deer population in areas of the US is one example, as is the elephant population in some reserves in Africa.

Landscapes

The extent to which natural lands are *connected* across a landscape is another concept of great importance in GI planning. Large areas of intact habitat could be considered connected if the habitat is *contiguous*, in other words, it is not fragmented by roads, buildings, and so on. However, even a fragmented landscape can maintain some degree of connectivity if there are pathways, or *corridors*, along which plants (seeds), insects, and larger animals can migrate from one larger patch of habitat to another. For example, a field or a well-buffered stream might provide adequate cover for many species to migrate between two habitat cores. A connected landscape is a healthier landscape; therefore, preserving the connectivity of natural lands can preserve critical ecosystem functions and should be a central objective of GI planning.

Key landscape elements central to GI planning

Core

A *core* (hub) is an area of relatively intact habitat that is sufficiently large to support more than one individual of a species. Accordingly, a small area of forest, say, might be a core for foxes, but not for bears. Thus, core often relates to the species under consideration. An American brown bear (grizzly bear) typically needs more than 250 square kilometers (~62,000 acres) of core habitat,[12] whereas a salamander may need only a small piece of land containing a vernal pool and adjacent uplands.

Thus, the amount of habitat a core requires will depend on the goals of a project. For example, some projects may choose several species that are of special importance and represent keystone, umbrella, indicator, or focal species that indicate the presence of other species. For example, the presence of a healthy population of mountain lions would indicate a healthy population of deer and other prey species.

Focal species are often used as surrogates because perfect information is not possible to obtain for every species in a habitat. The term *focal species* is used hereafter to refer to a surrogate species or a group of surrogate species used for this purpose. The habitats for these focal species are mapped, and the results are used as surrogates for other species when making land-use decisions. This method also brings some consistency when deciding on the size of cores—they are those in which the focal species can thrive.

Consider that the greater the number of species present and the greater the diversity of habitats, the more important it is to conserve a core. A key concept to understand when mapping cores is the notion of a two-part core composed of an *interior zone* and an *edge zone*. The interior zone is undisturbed habitat, whereas the edge zone surrounding it absorbs impacts from adjacent, nonnatural land uses (e.g., an urban area, road, or crop field).

The size and shape of a core can determine the amount of interior habitat. Credit: Green Infrastructure Center Inc.

In general, an edge zone is determined to be 100 yards (300 feet) deep, but this is a simplification of reality, because different types of impacts vary in magnitude. For example, soil moisture may be affected only 100 yards into a woodland, whereas the noise from a road might have significant effect 200 yards into the interior, and an invasive species may affect an area 500 yards from the edge. Nonetheless, *a defining characteristic of a core is the presence of a significant amount of interior habitat, which is free from edge effects*.[13]

Mapping core habitat has two common approaches. One approach determines the needs of a particular focal species and maps a habitat suitable for it. Expert knowledge is required to do this, and a team of conservation biologists commonly provides input and determines mapping rules for GI projects. Qualified scientists can address topics such as the forage, migration, and reproduction areas for a focal specie(s) in a study area.

A second approach involves taking an ecosystem-based view. This approach typically maps natural lands in combination with human modifications to determine an estimate of *ecological integrity*. Thus, habitat cores are defined as lands with minimal human impact (interior habitat). These areas are mapped under the assumption that they are a more suitable habitat for native species (e.g., pristine forest is more suitable than a parking lot). This approach provides a map of habitat cores.

A series of metrics can be calculated for each core (stored in an attribute table in an ArcGIS™ feature class). For example, how large is the core? Does it have any endemic species inside? Does the core have diverse soil types and topography?

These metrics are useful for further filtering the cores to estimate ecological integrity and prioritize conservation. Perhaps the single best predictor of ecological integrity is the size of a core. Larger cores tend to support a greater diversity of species and a greater number of habitat types. These cores are more resilient (less vulnerable to disturbances), support larger species population, and are more effective at protecting watersheds and aquifers.[14]

The Intact Habitat Cores dataset described in this book uses a minimum core size of 100 acres, which is the minimum recommended size for protecting many interior-dwelling birds in the eastern US.[15] This means that a patch of intact natural land must have at least 100 acres of interior habitat to qualify as a core. Larger cores may be needed to protect the best natural assets in a region, especially in the western US, where many species, such as mountain lions, have large ranges.

The process of mapping habitat cores and their associated metrics provides maps and data that are fundamental to constructing a GI plan. Chapter 5, Making asset maps, discusses how additional information can be overlaid on this base, such as hiking trails and groundwater recharge areas, to provide a more complete picture of the *multiple benefits* provided by these lands—a fundamental principle of GI planning.

Edge

An *edge* is the transitional boundary of a core, where the vegetation assemblage and structure differ markedly from the interior, such as forest edges. Many human alterations of the landscape produce effects that create an edge environment, including roads, housing developments, airfields, and mining operations. But natural phenomena also contribute to these effects, such as wind and sunlight penetration or the incursion of invasive species.

Effects of sun, wind, and human disturbance can cause impacts to the edge zone.
Credit: Green Infrastructure Center Inc.

Natural edge effects are categorized as either *biotic*, such as invasive species, or *abiotic*, such wind, temperature, and sunlight. The two often are related. For example, increased sunlight at a forest edge can increase temperature and decrease soil moisture, which can change the types of species that thrive, including invasive species.[16,17]

The magnitude of edge effects varies considerably and can depend on factors such as ecosystem type, edge type, and the species in question.[18,19] For example, increased wind speeds at the edge can disproportionately affect trees that have shallow roots and require a buffer, such as many wetland trees.[20] Although edge effects can vary greatly, especially biotic effects, an approach for estimating the extent of abiotic factors in forests is to multiply by 3 the average mature tree height.[21-23] For example, if a forest has an average tree height of 100 feet, the estimated edge zone would be roughly 300 feet from the outside.

Lands do not have to be urbanized to create edge effects; clear-cuts next to a forest can lead to loss of wildlife, exacerbate erosion, cause nutrient loss, and create additional runoff.[24] However, consider that a forest that has been clear-cut will likely regenerate with new growth, whereas a parking lot or building is a permanent hard edge.

Not all species are negatively affected by edge effects. Indeed, many species thrive in edge conditions, such as the indigo bunting or common rabbit. However, edge conditions fundamentally change the nature of their habitat.[25,26] Take, for example, one well-documented species that thrives in edge conditions—the white-tailed deer. The fragmentation of woodland by suburban lots has resulted in the overpopulation of deer, which has had many side effects, including overbrowsing of seedlings and saplings and overgrazing of forests. This fragmentation has changed the composition of species at the edges of these woodlands, often resulting in loss of species diversity and even encouraging other invasive species, because deer tend to avoid eating invasive vegetation, such as garlic mustard, Japanese barberry, and Japanese stiltgrass.[27,28]

Roads create more edge habitat and are ubiquitous, even in very rural landscapes. They can facilitate the spread of invasive species, and they create isolation and fragmentation. Even narrow, unpaved roads facilitate the movement of predators, changing an ecosystem's dynamics.[29,30] Roads create a variety of effects, with wide and more highly trafficked roads typically creating more impacts. They obviously create a barrier to animal movement, because many small forest mammals are less likely to cross even lightly traveled roads that are more than 18 feet wide and will rarely cross road zones (that include verges) greater than 90 feet wide. Many medium-sized mammals will cross road zones of that width, but large highway corridors form almost complete barriers to all animal movement. Large mammals will cross most roads, but crossing rates will be lower and will include higher rates of mortality.[31]

Roads create more edge habitat by bisecting natural areas. Credit: Green Infrastructure Center Inc.

Overall, higher road densities are correlated with animal avoidance of those areas.[32–34] Even birds, whose physical movement is not prevented by roads, can be affected by the noise and edge effects they create. For example, the noise effects of roads and urbanized areas can interfere with bird calls, which can affect mating, locating prey, and finding separated young.[35,36]

Corridor

Maintaining and enhancing connectivity across a landscape are key components of GI planning. To turn habitat cores into a network requires the connectivity provided by interspersed tracts of habitat that facilitate the movement of animals, plants, and pollinators between cores and prevent species within them from becoming isolated.

A *corridor* is a basically linear arrangement of habitat type or natural cover that connects cores and differs from adjacent land on either side, such as a well-buffered stream. Species use corridors to move between cores, which have a long history as a valuable conservation tool. The ability to move between cores creates healthier and more resilient landscapes for animals, plants, and pollinators.[37]

In relation to climate change, the maintenance and creation of a connected landscape is one of the mitigation techniques most frequently proposed by ecologists to allow species to adapt to developmental changes and remains an area of active research.[38,39] Cores that are connected tend to retain more native plant species than do isolated habitats.[40]

People may think of corridors as a landscape feature that supports the movement of large, terrestrial mammals. Although this is true, the term *corridor* as used in this book is defined broadly as a landscape feature that also supports insects (e.g., pollinator pathways), plants (e.g., seed dispersal), aquatic corridors, and seasonal migration corridors.

Migration corridors

The two main types of *seasonal migration* are *elevation migration* and *latitude migration*. Each type of migration relies on different types of corridors, which in turn serve as connective landscapes that facilitate the movement of species according to the seasons. For example, elevation migration includes the migration of elk on the Colorado Plateau of Arizona, New Mexico, Colorado, and Utah into the high valleys and mountain slopes during the summer months and their return to the low valleys in the fall rutting season, where they remain to overwinter. In the southeastern US, black bears also move to lower elevations to spend the winters and return to higher elevations in the summer. In the Northwest, grizzly bears migrate in reverse—moving up into the snow pack to build overwinter dens to give birth to their young and returning to lower elevations to forage in the spring.

Species also often move great distances along latitude migration corridors, crossing latitudes from season to season. In North America, numerous birds, including many hawk species, and insects, including the monarch butterfly, move across these corridors. At one time, bison and antelopes migrated this way, although movement of these large mammals is now severely constricted.

Elevation and latitude migrations result from a combination of factors, such as temperature, available food resources, rainfall, and protection of young from predators, although they all occur as seasonal movements. In both cases, major conflicts can arise between animals and humans. US Route 395 in California crosses the entire length of the seasonal elevation migration of mule deer and pronghorns from the eastern slopes of the Sierra Nevada mountain range to the Nevada desert, causing significant danger to both animals and drivers. In general, the movement of species cannot be ignored, because animals are not aware of traffic signs, property rights, and legal rulings. Even at a micro level, considering the behavior of local species will help prevent conflict.

Recolonization

Corridors allow species to recolonize isolated areas that may have lost their original population, which is crucial in fragmented landscapes. For example, if a species in a core area is in danger of becoming isolated because habitat conditions have become unsuitable, it is more likely to survive if it can use corridors to move to new habitats nearby.

Wider corridors are correlated with greater species abundance and diversity.[41] How wide is wide enough? The answer will depend on the ecosystem and the species, but generally wider is better. Ideally, an interior habitat needs to be present for it to be a corridor. In forested landscapes, for example, the connections should be at least 900 feet wide. The corridor should include a central, 100-foot-wide pathway of interior habitat and a 100-foot edge on either side to protect safe passage and buffer against human intrusion and invasive species. Streams are natural corridors, and the width of the vegetative corridor on either side should reflect the stream order (i.e., larger streams need wider forested buffers).

Larger and more complex networks of interconnected corridors and cores are more likely to have more overall species diversity than are smaller networks, and functioning ecosystems can be maintained amidst a changing landscape. Remember, it is far easier to maintain an existing connection than it is to restore one, although corridors often can be restored. However, the best and most critical existing connections must be identified, even if they are in relatively good shape, so they can be included in land-use planning processes and protected against future development pressures. If a corridor is not identified, one cannot determine whether it is at risk of being lost or degraded.

Given the critical importance of connective corridors to an integrated landscape, planners and decision makers should also implement policies that will identify where corridors are lacking and plan to reestablish them. Existing corridors that lack adequate interiors can also be widened. Often, local landowners are willing to work with planners, especially if given incentives to preserve and protect their natural corridors. For example, some local governments purchase conservation easements to take up development rights, leaving the land in private ownership while compensating landowners for giving up their rights to develop it.

Before

After

Corridors are natural, linear patches of natural land cover that connect core habitats. Ideally, an interior habitat is present in a corridor. Corridors can be created or widened by restoring habitat. Credit: Green Infrastructure Center Inc.

Stepping stone

Interspersed among most networks of core areas and corridors, certain smaller areas act as stepping stones. Even in the absence of intact corridors, these smaller habitat patches can help species disperse across a landscape. A stepping stone tends to be a smaller area of intact habitat that may not be large enough to sustain a species on its own but is vital to a population's success over the network as a whole (its *metapopulation*). Without these stepping stones, cores can become increasingly isolated as species lose diversity on *habitat islands*. However, if those species maintain a stable population somewhere (e.g., in a large core), they may maintain their populations by dispersing to the isolated and fragmented patches, which may offer shelter and food, if not a sustainable number of breeding individuals.

Even in the absence of intact corridors, stepping stones can provide a way for species to disperse across a landscape. Credit: Green Infrastructure Center Inc.

> ### A fragmented network consists of three types of habitat
>
> Within the context of a human-influenced and fragmented landscape, a sustainable network of (relatively) thriving species consists of three kinds of habitats:
>
> - Intact core habitats
> - The habitat corridors that connect them
> - Those smaller areas of habitat that serve as stepping stones between larger cores

Landscape metrics

> *Mosaic is a term often used to describe the network of cores, corridors, and stepping stones set within the context of all other patches (land cover types) that form a landscape in its entirety.*

The landscape metrics for GI planning are indices developed for categorical map patterns; they are algorithms that quantify specific spatial characteristics of the cores, corridors, and stepping stones, as well as their associated values, for an entire landscape. There are two types of landscape metrics: those that do not reference spatial attributes, and those that do. These metrics are landscape *composition*, which includes the proportion of a landscape in each patch type, together with individual core richness, core evenness, and core diversity, as just one of the possible ways to quantify composition. They also include such concepts as isolation, contrast, contagion, dispersal, and interspersion (see sidebars).

Principle measures of landscape composition

The principle measures of landscape composition, as adapted to GI planning, include the following:

- **Proportional abundance:** The proportional abundance of each habitat type relative to the entire landscape under consideration.

- **Richness:** The number of different habitat types.

- **Evenness:** The relative abundance of each habitat type: Is there a relative dominance of one type or relative equality of types? Evenness is usually reported as a function of the maximum diversity possible for a given richness; it is given as 1 when the mosaic is perfectly diverse and approaches 0 as evenness decreases.

- **Diversity:** A composite measure of richness and evenness; it can be computed in a variety of forms,[42] depending on the relative emphasis placed on these two components.

Some other technical terms

Here are some other useful technical terms relating to GI landscape planning:

- **Spatial configuration** refers to the spatial character and arrangement, position, or orientation of patches within a landscape. Some aspects of configuration, such as *core isolation* or *core contagion*, measure the placement of cores, corridors, and stepping stones relative to each other or to other features of interest. This metric recognizes that the ecological properties of a core are influenced by its surrounding network (mosaic), as well as by edge effects, core size, and core shape.

- **Contrast** refers to the relative difference between a core and its neighboring patch (land cover) types. For example, mature forest next to younger forest will have a lower contrast edge than mature forest adjacent to a housing development. Cores with high contrast to their neighboring land covers tend to greater isolation.

- **Contagion** measures the extent to which patch (land cover) types clump together. *Interspersion*, conversely, refers to the intermixing of patches of different types. All other things being equal, a landscape in which the patch types are well interspersed will have lower contagion than will a landscape in which patch types are poorly interspersed.

Modeling connectivity across a landscape using GIS

GI planning essentially models the connectivity of habitats across a landscape. It maps and assesses patches of habitat in terms of their viability and quality and the extent to which they form an adequate network, across which species can move and thrive. Within that context, GI planning also can consider human needs, such as clean water, forestry products, recreation, and tourism. It includes established datasets and models of future human developments, such as roads, business parks, quarries, and housing developments. It includes zoning and highlights aspects of comprehensive plans, as well as community input. It then creates a series of map overlays that highlight the relationships between the habitat network and these various other factors. Thus, GI planning is a tool for planners and decision makers to identify potential conflicts, opportunities, and impacts of those decisions on the environment.

Cost-distance analysis

The most common approach for modeling connectivity across a landscape uses *cost-distance* analysis. In this approach, the surface of the earth is represented as a raster dataset. Each cell in the raster has a numeric value, but instead of representing elevation (as in a digital elevation model) or a type of land cover, the value represents the *level of effort required to move across that cell*. However, the values may not consist of fixed units of measurement (e.g., feet-per-minute), but rather may have meaning only *in relation to each other*. For example, if modeling a species' ability to walk across the landscape, one might assign a high value to any cell with a steep slope and a low value to flat land, a high value to a wide river and a low value to a stream, and a high value to a freeway and a low value to a dirt road.

> *As high value represents more effort required to move across a cell, in identifying corridors for species movement, planners should connect cells of the lowest possible values, which will tend to follow lines of consistent elevation, follow the edges of watercourses rather than cross them, and avoid wide, busy roads.*

This raster is called a *cost surface*, which can be created by using multiple datasets that represent features that impede or facilitate the cost of travel across a landscape. For example, *land cover* is one of the most commonly used factors when modeling wildlife corridors. A land cover raster can be reclassified to represent the effort it takes to move through types of cover. For example, natural areas typically have a lower cost, whereas urbanized areas have a higher cost.

Roads are another common dataset used in this model, because they represent one of the most ubiquitous and insidious barriers to wildlife movement. A GIS roads layer can be converted to a raster dataset and the cells assigned a cost (impedance) value. This new raster is then merged with the reclassified land cover raster to create a simple cost surface.

A conceptual 7-by-7 cell cost raster. The number in each cell indicates the cell's value, which represents the cost of traveling across the cell. In this simplistic example, green cells represent forest, brown cells represent agriculture, gray cells represent urban areas, and blue cells represent water. The cost values show that forest (natural) areas have the least cost to travel across, whereas agricultural and urban areas have greater costs. Water, in this example, effectively serves as an impassable barrier. Credit: Green Infrastructure Center Inc.

For more background on cost-distance modeling, see **The Esri Guide to GIS Analysis, Volume 3: Modeling Suitability, Movement, and Interaction (Esri Press, 2012).**

To model a corridor between two habitat cores, use a cost surface raster in conjunction with the two cores, which serve as *sources*, to model the path of least resistance (*least-cost path*) between them.

Concepts of scale

Scale is a fundamental topic in landscape ecology. In general, truly informed land management planning cannot be conducted exclusively at the scale of individual habitat units or isolated local sites. Expansive phenomena, such as climate change, biodiversity protection, and acid rain, have forced the environmental community to think at grander, and often more challenging, scales.[43] However, for most planners, the local scale is the most obvious and immediate—and the one that needs action. One of the major problems is that expansive phenomena that affect species at a local scale cannot always be extrapolated from the larger scale.[44] The answer to this problem lies, in part, in taking a multiple-scale approach. Many datasets bridge the divide between national or even global trends, and how they affect regional or statewide landscapes. Data available from other local sources such as universities or conservation organizations can be utilized to inform climate projections, analyses of species diversity, endangered species, distinct local habitat types, and so on. For example, in the southwestern US, universities and states have data on rainfall projections; in the Midwest, they have projections on aquifer usage; and in the Southeast, they have projections on impacts from hurricanes and future sea level rise.

Geographic extent

Before any analysis in GIS is performed, two other factors need to be defined: geographic extent and spatial resolution.

Geographic extent defines the limits of an area of interest. Analysis extents can vary greatly, but common extents for GI planning include cities, counties, multicounty regions, and states. They also include specific habitats, such as an individual forest, a watershed, or a park. Watersheds make ideal extents—units of analysis—for GI planning, but they may cross administrative boundaries (city, county, state), which makes project coordination more difficult. You will need to obtain data from all the jurisdictions within the watershed.

Most important, the geographic extent of a GI planning project depends on the project's goals. If the primary goal is to improve water quality, a watershed may be the ideal planning extent for the map; if the goal is to control an invasive species, a forest-wide extent may be required; if the goal is to provide greater tourist opportunities within a county, then that county's boundaries will likely be the extent; if it is a specific project to beautify entrance corridors, the extent will be limited accordingly along the viewshed of the highway.

Spatial resolution

Spatial resolution defines the smallest polygon or raster cell size of a dataset. Spatial resolution is also referred to as the *grain* of the data. For example, a raster land cover dataset that is mapped at 1-meter resolution contains finer-grained data than a land cover dataset mapped at 30-meter resolution. A key question when considering spatial resolution is, What is the smallest feature that can be distinguished? The 30-meter dataset is useful when analyzing broad landscape patterns for a multistate or multicounty region, whereas urban areas require 1-meter resolution datasets.

Determining extent and resolution

How does one define the appropriate geographic extent and spatial resolution for a GI project? In practice, this is often constrained by various factors. A county planner may not be able to justify working at a regional scale. However, she may choose to map all watersheds across which their county boundaries extend, including water courses not within their jurisdiction. Doing so allows the planner to see patterns and connections that would not be obvious if the maps were cut off at the county's boundaries. Of course, a county planner cannot set policy for neighboring jurisdictions, which is why cross-county collaboration is extremely useful for a successful plan. Ideally, multilocality issues would be identified in a GI planning project, and all impacted localities would begin collaborating as early as possible.

Cross-jurisdiction cooperation can take time to build, because so many decision makers have to agree to a project's goals and extent, but many GI plans necessitate cross-boundary analysis and collaboration. For instance, most cities' water supplies are located beyond city limits; city outflows of sewage and stormwater affect downstream counties and water bodies; cities place great burdens on surrounding counties' natural features and leisure opportunities; and road networks require multijurisdictional planning.

Spatial resolution is often limited by data availability, particularly land cover data. Creating new land cover data can be costly, as it requires purchasing and classifying aerial imagery. Fortunately, there are publicly available land cover datasets, such as the National Land Cover Database. The NLCD provides national coverage at a resolution appropriate for landscape-scale GI planning.

One rule of thumb for determining the appropriate resolution for raster data is *Resolution in meters = Scale factor × 0.0005*.[45] Thus, if one has a county scale map at 1:100,000, the approximate cell size would be 50 meters. The NLCD, at a resolution of 30 meters, would meet the requirements using this rule of thumb.

It is important to keep in mind that the patterns and interactions being modeled can be highly dependent on the scale at which analysis is to be performed.

There is no ideal scale of analysis for all species

A bald eagle functions at a different scale than a mouse, which functions at a different scale than a beetle. Even if we are interested only in bald eagles, there is no single scale that works best for assessing their habitat. When an eagle's three primary habitat requirements (access to foraging areas, available nesting sites, and isolation from human disturbance) are assessed individually, they each have a different ideal geographic scale and spatial resolution— the scale at which they respond to; statistically, this is *habitat availability* plotted against *habitat use*.[46]

There is no ideal scale for assessing habitat fragmentation

The metrics used to quantify landscape fragmentation are affected by the raster resolution used in the analysis. Some metrics, such as the total number of habitat cores on a landscape, will vary in a predictable way. Others will vary unpredictably.[47] One such unpredictable variable is the *contagion metric*—a measure of how aggregated the patch types are. This has higher values for a landscape with fewer large patches and lower values for a landscape with a proliferation of smaller patches.

Unfortunately for planners, there is no single analysis scale that works for all ecological phenomena.[48] The complex interaction of processes occurring at difference scales of space and time, coupled with the fact that landscape ecology is often constrained by observational rather than experimental research design, can mean that definitive rules are hard to come by. However, there is a long history of using models to enrich our understanding of these phenomena and to support better land management decisions.

Modeling large landscapes

Both nationwide and regional modeling can play a significant role in local models, given the wider context of many ecological, environmental, and cultural features of the landscape, for example, the migration routes of butterflies, cranes, or elk; the context of a battlefield; or the context of a county's stretch of river within an entire watershed.

Modeling accumulated changes

If natural assets are not included in a model of future developments in a locality, if there is no baseline inventory of those assets created and mapped, then, over the years, data concerning those assets will become inaccurate.

When considering what the best natural assets are, a lack of action commonly leads to the degradation or loss of these assets. This is because numerous small developments are taking place over time without accounting for their collective implications. This often leads to effects such as deforestation, loss of agricultural potential, loss of environmental buffers for hazards, and unhindered growth.

The piecemeal and often haphazard nature of urbanization is one reason why "death by a thousand cuts" is an analogy frequently used when describing the worldwide problem of loss of habitat, cultural and scenic assets, agricultural areas, and protection from hazards. Buildings are built and natural landscapes are cleared of vegetation without understanding the impacts of breaking the landscape into smaller, disconnected units. Smaller units—patches—lead to landscapes with lower value for a community and less habitat for wildlife and other native species.

The ecology of GI landscape planning

Nationwide modeling

The availability of nationwide data and modern computing power have allowed for modeling of very large landscapes, even at relatively fine-grained spatial resolutions. In 2012, David Theobald and a team of researchers had modeled potential corridors for the nation using the concept of *landscape permeability*. This approach also employed the concept of cost distance, but differed from the later study, which used large protected areas as cores (origins/destinations in the network).[49] Instead, the entire landscape (lower 48 US states, in this case) was treated as a continuous gradient of resistance, which was also based on a cell's naturalness (using a 270-meter resolution cost surface based on levels of human modification). The resulting corridors reflected areas that were used more often as pathways when considering the flow across the entire landscape (rather than between two specific cores).

In 2016, R. Belote and colleagues modeled potential corridors between large protected areas in the US. He employed the concepts of *cost distance* and defined cores as "large protected areas," a different definition from that given here, because a habitat's protection status does not determine whether it qualifies as a core. To find these *paths of least resistance*, resistance values were estimated by creating cost rasters at a 1-kilometer resolution based on *naturalness*. This term was defined according to how heavily modified an area was by human intervention.[50]

How the national model was created

Habitat cores are intact areas of natural land cover at least 100 acres in size and at least 200 meters wide. They are derived from the best available national data to create complete coverage for the contiguous US. Potential core areas are selected from the National Land Cover Database (2011) using land cover categories not containing the word *developed* or those categories associated with agriculture uses (*crop*, *hay*, and *pasture lands*). The resulting areas are tested for size and width and then converted into unique polygons. These polygons are then overlaid with a diverse assortment of physiographic, biological, and hydrographic layers to calculate metrics that describe each core. These metrics are the foundation for assessing what makes each core unique and whether a core should be a conservation priority. (See Appendix B for a full list of ancillary data used and core metrics.)

How to access the data

You can access the data for the national GI model in several ways:

- **Download the cores data by state:** The raw data (all habitat cores and their attributes) are available by state, downloadable through ArcGIS Online. This method provides map packages that include the habitat cores data, as well as a variety of source information that was used to generate the cores and their associated metrics.

- **Use the Esri GI apps:** Several web apps make it easy to download cores for specific areas, as well as perform custom filters and rankings.

- **Download the toolbox:** For more experienced GIS users, a toolbox is available that lets users run the habitat core model themselves.

In choosing how to access the model and source data, consider familiarity with ArcGIS software, the intended use of the data, and access to a desktop application (ArcMap™, ArcGIS Pro™).

For example, a GIS analyst who routinely uses ArcGIS Pro may find it most effective to download the map package for a specific state. A county planner who does not have GIS software installed on the desktop may find it easiest to use the online applications to view and extract the habitat cores for his or her county. The planner might then use ArcGIS Online to make several web maps that overlay other key county datasets, which would be shared with the rest of the department.

Green infrastructure for the US: Esri's national green infrastructure model

The national GI model provides data that are locally relevant, adaptable, and easy to use:

- **Locally Relevant:** The data are usable for communities, not just for large regions. To accomplish this, the model uses the National Landcover Database at its original resolution—30 meters (this is the spatial resolution of the model as defined in the Scale section in this chapter). The NLCD is the finest resolution land cover dataset available for the entire contiguous US. Other datasets are used in the analysis, but the NLCD is the fundamental piece of the analysis. As such, its resolution defines much of the resulting information.

- **Adaptable:** The model is designed as a general-purpose product that users can tailor to more specific objectives. More than 40 metrics are calculated for each core to facilitate filtering and prioritizing based on local objectives. The model estimates ecological integrity rather than suitability for a species. For example, the model is not species specific, but if one is specifically interested in mountain lion habitat, the data can be queried using its wealth of landscape metrics to easily jumpstart the process. Likewise, if pronghorns (*Antilocapra americana*) are of special concern, users can perform a separate query to identify suitable habitat. You should include local and expert knowledge at this point. If the model was designed as a species-specific product, it would be of limited use, because landscapes and conservation priorities vary tremendously throughout the country.

- **Easy to Use:** Users can access the model in a variety of ways, according to their technical ability. The raw data can be downloaded and manipulated in ArcGIS Desktop, or they can be accessed and manipulated using online apps. Habitat cores can be ranked using different scenarios and overlaid with local data without ever leaving ArcGIS Online.

The results are intended to promote GI planning by making it as easy as possible to get started.

Regional modeling

There have been a variety of efforts to model ecological assets and connectivity at regional, multistate scales. Because these types of projects cross multiple states and jurisdictions, stakeholder cooperation is a must. In the US, the Department of the Interior launched a network of 22 Landscape Conservation Cooperatives (LCCs) designed to address landscape scale issues such as climate change and habitat connectivity. Each LCC has created a collaborative partnership between scientists and resource managers from government authorities, nongovernmental organizations, tribes and First Nations, universities, and other stakeholders.

Begun in 2004, *Two Countries, One Forest* is a Canadian–US collaboration focused on protecting, conserving, and restoring forests and natural heritage sites from New York to Nova Scotia, encompassing the Northern Appalachian/Acadian ecoregion. As part of this initiative, many models have been built to estimate ecological integrity, connectivity, human footprint, and resilience to climate change. A digital atlas offers more information to practitioners involved with land-use planning and decision making at **https://2c1forest.databasin.org/**.

The Great Northern Landscape Conservation Cooperative (GNLCC) Ecological Connectivity Project is addressing ecological connectivity from British Columbia in Canada to western Wyoming in the US. The group uses both top-down and bottom-up strategies to foster collaboration among practitioners and create large-scale landscape models. Their digital atlas is at **https://databasin.org/galleries/5c51bde995e84581b444b9bec7942b43**.

In the Southeast, the South Atlantic Conservation Blueprint models a variety of ecological indicators from Virginia to Florida. Corridors that link coastal and inland areas and span climate change gradients have also been modeled using the concept of least-cost paths. See **http://www.southatlanticlcc.org/blueprint/**.

State models: Florida

Some states have created statewide models of their landscapes using the principles of habitat cores and corridors. In Florida, one of the first states to implement a statewide landscape model, GI planners proposed two networks:

- An **ecological network**, consisting of ecological hubs, linkages, and sites along rivers and coastlines and across watersheds
- A **recreational/cultural network**, with trail corridors connecting parks, urban areas, working landscapes, and cultural/historic sites

To design the green network, Tom Hoctor, Margaret Carr, and Paul Zwick led a research team at the University of Florida (UF) to create a network of cores and corridors resulting in the Florida Ecological Greenways Network design. They used a raster format with Esri's grid model within ArcInfo software. To allow for faster data processing, it used 180-meter resolution for analysis, which was later improved to a 30-meter cell size for finer analysis, as processing speeds improved with better technology.

Working with a team of technical advisors, they identified areas of ecological significance by assembling and classifying data layers, including strategic habitat conservation areas, priority natural communities, conservation lands, roadless areas, and significant aquatic systems. They created rankings for each of the layers and combined all areas with primary ecological significance into a single layer titled "Priority Ecological Areas." The map's purpose was to help the state identify key areas for both recreation and ecological protection.

In 2001, these UF researchers applied a similar modeling approach to create a multistate regional map for the Southeastern Ecological Framework; a large-scale (90-meter) map of eight southeastern states funded by the US Environmental Protection Agency Region IV. Data on which the work was built were acquired for the entire region and from individual states. The purpose was to develop a GIS database of conservation priorities to facilitate coordination between agencies using the best available data and the goals of conservation biology "to understand natural ecological systems well enough to maintain their diversity in the face of an exploding human population."[51] They also used principles from the field of landscape ecology.

The Florida Ecological Greenways Network continues to inform state planning, land acquisition, and education. For more, see chapter 8. This demonstrates a key best practice: utilizing GI planning as an ongoing process. GI data and maps are living documents that should evolve as land use changes and as new information, computer software, and science become available.

State models: Maryland

Maryland's GreenPrint is a program designed to identify and protect the state's most valuable ecological lands. Landscape-scale habitat modeling was a key component. It identified cores and corridors and showed the relative ecological importance of every parcel of land in the state. At the county level, Prince George's County adopted the model and utilizes it to conserve GI locally. Its first plan identified 92 strategies for conserving or expanding the county's GI. The updated plan will reflect new priorities and new data. This work demonstrates how broader GI plans can be included in local-scale planning.

State models: Virginia

In 2007, the Virginia Department of Conservation's Natural Heritage Program developed its statewide Virginia Natural Landscape Assessment. This model ranks habitat cores using a variety of factors, such as size, depth, isolation, topographic diversity, wetland variety, biodiversity, and length of interior streams. It has been used by regions, cities, and towns to prioritize natural assets.

To ensure the model was used locally and regionally, the Green Infrastructure Center stepped forward to ensure regional and local governments knew how to use the model for planning. Beginning in 2006, it tested the model's application at a variety of regional and local scales, incorporating local data to make the model useful at the local level. It produced locally relevant GI network maps and themed overlay maps to highlight key issues for each locality, such as water resources, recreation, heritage, and agriculture. Since 2006, it has conducted more than 40 projects in Virginia's regions, counties, cities, and towns to help them develop their GI plans.

The Virginia Natural Landscape Assessment identifies and ranks core habitat, corridors, and natural landscape blocks. Credit: Virginia Natural Landscape Assessment model by the Division of Natural Heritage, Department of Conservation and Recreation.

This demonstrates another key component of GI planning: there should be a way for communities to utilize models to inform their planning efforts. In the case of the Virginia model, the Green Infrastructure Center was able to provide this technical assistance. Since then, the Green Infrastructure Center has branched out into other states, where it has provided extensive consultation and teaching and training services and built statewide habitat models. To date, it has created such models for New York, Arkansas, and South Carolina. The South Carolina model formed the prototype for the Esri green infrastructure national model.

Other state models

Other states have created landscape models that employ similar concepts and methods:

- **Washington**—Washington Connected Landscapes Project: Statewide Analysis
- **California**—California Essential Habitat Connectivity Project
- **Montana**—Montana Connectivity Project
- **Arizona**—Arizona Landscape Integrity and Wildlife Connectivity Assessment
- **Massachusetts**—BioMap2: Conserving the Biodiversity of Massachusetts in a Changing World

Montana Connectivity Project
All General

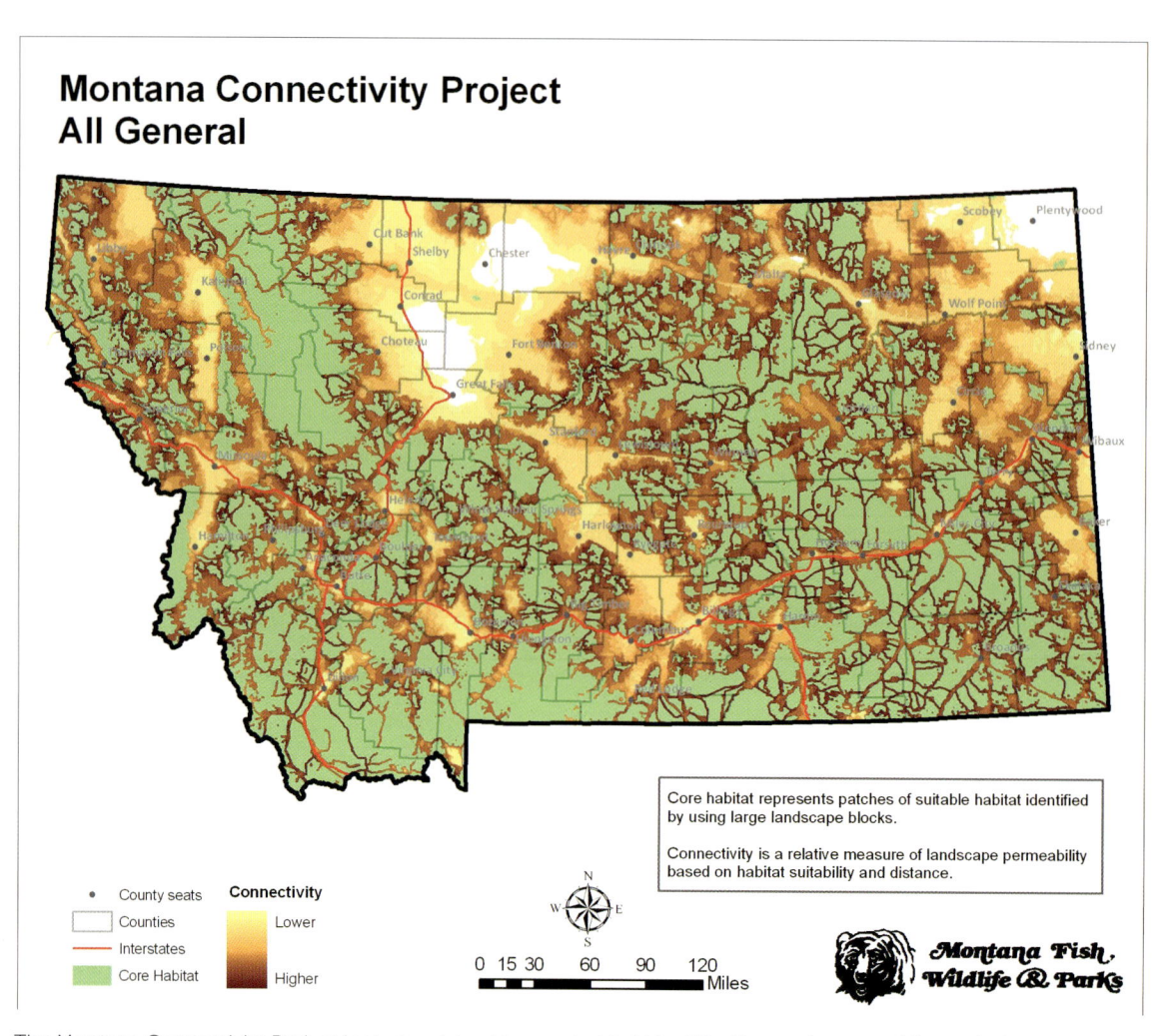

The Montana Connectivity Project includes statewide core habitat identification and connectivity analysis at several levels (species, species guild, and landscape blocks). Credit: *Montana Connectivity Project—A Statewide Analysis.* Montana Fish, Wildlife, and Parks, 2011.

Modelers and planners should check with their state's conservation division or natural heritage staff to learn what data and tools have been created. These tools may be used to increase the accuracy or specificity of the national model.

In this chapter, we explained the key concepts of habitat connectivity and modeling—core, edge, corridor, migration routes, stepping stones, and connectivity modeling. We also explained the landscape metrics used to evaluate the habitat's ecological integrity and what scales of analysis are appropriate based on your goals. Finally, we described the Esri GI model, which is based on this approach to habitat modeling. Next, chapter 3 will discuss the goal-setting process that informs the entire GI planning effort.

Notes

1. Forman, Richard T. T., and M. Godron, *Landscape Ecology* (New York: John Wiley & Sons, 1986).
2. Harris, L. D., "Special visual presentation. Landscape linkages: The dispersal corridor approach to wildlife conservation." *Transactions of the 53rd North American Wildlife and Natural Resources Conference* (Louisville, KY, 1988).
3. Beier, Paul, and Reed F. Noss, "Do habitat corridors provide connectivity?" *Conservation Biology* 12, no. 6 (1998): 1241–1252.
4. Firehock, Karen E., *Evaluating and Conserving Green Infrastructure across the Landscape: A Practitioner's Guide* (Charlottesville, VA: The Green Infrastructure Center Inc., 2012).
5. Tewksbury, Joshua J., Douglas J. Levey, Nick M. Haddad, Sarah Sargent, John L. Orrock, Aimee Weldon, Brent J. Danielson, et al., "Corridors affect plants, animals, and their interactions in fragmented landscapes." *Proceedings of the National Academy of Sciences* 99, no. 20 (2002): 12923–12926.
6. Parrish, Jeffrey D., David P. Braun, and Robert S. Unnasch, "Are we conserving what we say we are? Measuring ecological integrity within protected areas." *BioScience* 53, no. 9 (2003): 851–860.
7. Hanski, Ilkka, "Predictive and practical metapopulation models: The incidence function approach." In *Spatial Ecology: The Role of Space in Population Dynamics and Interspecific Interactions*, ed. David Tilman and Peter Kareiva (Princeton, NJ: Princeton University Press, 1997), 21–25.
8. MacArthur, Robert H., and Edward O. Wilson, *The Theory of Island Biogeography* (Princeton, NJ: Princeton University Press, 2016).
9. Harris, Larry D., *The Fragmented Forest: Island Biogeography Theory and the Preservation of Biotic Diversity* (Chicago: University of Chicago Press, 1984).
10. Sorrell, J., "Using geographic information systems to evaluate forest fragmentation and identify wildlife corridor opportunities in the Cataraqui watershed" (Faculty of Environmental Studies, York University, Ont., Canada, 1997).
11. Haddad, Nick M., Lars A. Brudvig, Jean Clobert, Kendi F. Davies, Andrew Gonzalez, Robert D. Holt, Thomas E. Lovejoy, et al., "Habitat fragmentation and its lasting impact on Earth's ecosystems." *Science Advances* 1, no. 2 (2015): e1500052.
12. Walker, Richard, and Lance Craighead, "Analyzing wildlife movement corridors in Montana using GIS." *Proceedings of First Annual James Reserve Conference, July 5–7, 1997.*
13. Zipperer, Wayne C., "Deforestation patterns and their effects on forest patches." *Landscape Ecology* 8, no. 3 (1993): 177–184.
14. Forman and Godron, *Landscape Ecology*.
15. Ibid.
16. Ibid.
17. Brown, M. T., J. M. Schaefer, and K. H. Brandt. "Buffer Zones for Water, Wetlands, and Wildlife in East Central Florida. Final Report, Florida Agricultural Experimental Station." *Center for Wetlands, University of Florida, Gainesville, FL* (1990).
18. Ewers, Robert M., and Raphael K. Didham, "Confounding factors in the detection of species responses to habitat fragmentation." *Biological Reviews* 81, no. 1 (2006): 117–142.
19. Alignier, Audrey, and Marc Deconchat, "Patterns of forest vegetation responses to edge effect as revealed by a continuous approach." *Annals of Forest Science* 70, no. 6 (2013): 601–609.
20. Brown, Schaefer, and Brandt, *Buffer Zones for Water*.
21. Harris, *The Fragmented Forest*.
22. Brown, Schaefer, and Brandt, *Buffer Zones for Water*.

23. Kapos, V., G. Ganade, E. Matsui, and R. L. Victoria. "α13C as an indicator of edge effects in tropical rainforest reserves." *Journal of Ecology* (1993): 425–443.

24. Harris, *The Fragmented Forest*.

25. Brown, Schaefer, and Brandt, *Buffer Zones for Water*.

26. Kapos et al., "α13C as an indicator of edge effects."

27. Averill, Kristine M., David A. Mortensen, Erica A. H. Smithwick, and Eric Post, "Deer feeding selectivity for invasive plants." *Biological Invasions* 18, no. 5 (2016): 1247–1263.

28. Forman, Richard T.T., and Anna M. Hersperger, *Road Ecology and Road Density in Different Landscapes, with International Planning and Mitigation Solutions*. Publication No. FHWA-PD-96-041 (Tallahassee: Florida Department of Transportation, 1996).

29. Ibid.

30. Gucinski, H., M. Furniss, R. Ziermer, and M. Brookes, *Forest Service Roads: A Synthesis of Scientific Information*. General Technical Report PNW-GTR-509.1 (Portland, OR: USDA Forest Service, Pacific Northwest Research Station, 2001).

31. Forman and Hersperger, *Road Ecology and Road Density*.

32. Ibid.

33. Gucinski et al., *Forest Service Roads*.

34. Mladenoff, David J., Theodore A. Sickley, Robert G. Haight, and Adrian P. Wydeven, "A regional landscape analysis and prediction of favorable gray wolf habitat in the northern Great Lakes region." *Conservation Biology* 9, no. 2 (1995): 279–294.

35. Brown, Schaefer, and Brandt, *Buffer Zones for Water.*.

36. Yahner, Richard H. "Changes in wildlife communities near edges." *Conservation Biology* 2, no. 4 (1988): 333–339.

37. Tewksbury et al., "Corridors affect plants, animals."

38. Heller, Nicole E., and Erika S. Zavaleta, "Biodiversity management in the face of climate change: A review of 22 years of recommendations." *Biological Conservation* 142, no. 1 (2009): 14–32.

39. Beier, Paul, "Conceptualizing and designing corridors for climate change." *Ecological Restoration* 30, no. 4 (2012): 312–319.

40. Damschen, Ellen I., Nick M. Haddad, John L. Orrock, Joshua J. Tewksbury, and Douglas J. Levey, "Corridors increase plant species richness at large scales." *Science* 313, no. 5791 (2006): 1284–1286.

41. Lindenmayer, David B., and Jerry F. Franklin, *Conserving Forest Biodiversity: A Comprehensive Multiscaled Approach* (Washington, DC: Island Press, 2002).

42. Shannon, C.E., and W. Weaver. *The Mathematical Theory of Communication*. Urbana: University of Illinois Press (1949). Simpson, Edward H. "Measurement of diversity." *Nature* 163, no. 4148 (1949): 688.

43. Turner, Monica G., Robert H. Gardner, and Robert V. O'Neill, *Landscape Ecology in Theory and Practice*, vol. 401 (New York: Springer, 2001).

44. Schneider, David C., "The rise of the concept of scale in ecology: The concept of scale is evolving from verbal expression to quantitative expression." *BioScience* 51, no. 7 (2001): 545–553.

45. Gucinski et al., *Forest Service Roads*.

46. Thompson, Craig M., and Kevin McGarigal, "The influence of research scale on bald eagle habitat selection along the lower Hudson River, New York (USA)." *Landscape Ecology* 17, no. 6 (2002): 569–586.

47. Mladenoff et al., "A regional landscape analysis."

48. Levin, Simon A., "The problem of pattern and scale in ecology: The Robert H. MacArthur Award Lecture." *Ecology* 73, no. 6 (1992): 1943–1967.

49. Theobald, David M., Sarah E. Reed, Kenyon Fields, and Michael Soule, "Connecting natural landscapes using a landscape permeability model to prioritize conservation activities in the United States." *Conservation Letters* 5, no. 2 (2012): 123–133.

50. Belote, R. Travis, Matthew S. Dietz, Brad H. McRae, David M. Theobald, Meredith L. McClure, G. Hugh Irwin, Peter S. McKinley, Josh A. Gage, and Gregory H. Aplet, "Identifying corridors among large protected areas in the United States." *PLOS ONE* 11, no. 4 (2016): e0154223.

51. Meffe, G. K., and C. R. Carroll, "Conservation reserves in heterogeneous landscapes." In *Principles of Conservation Biology*, 2nd ed. (Sunderland, MA: Sinauer Associates, 1997), 305–343.

Chapter 3

The six-step process

A six-step process is recommended to create green infrastructure (GI) maps and strategies. This step process was developed by the Green Infrastructure Center (GIC) to provide communities with a logical and iterative process for arriving at a strategic plan for conserving and restoring habitat cores and related natural and cultural assets. Based on more than 10 years of GI planning field tests conducted by the Green Infrastructure Center in wild, rural, suburban, and urban environments, the step process is appropriate for varied development patterns, from densely to sparsely developed landscapes.

The process

This step process provides a framework for thinking about GI planning and what strategic actions can be taken to conserve or restore GI. The steps help users and planners identify, assess, evaluate, and prioritize GI assets. Although you can use the Esri GI model to generate a map, that map itself is not a plan—it requires accompanying strategies. Most important, you should create a map with clear goals in mind, so that it becomes a tool to realize key strategies. This chapter explains how you can set goals to inform GI maps and related strategies to meet real needs on the ground.

GI planning has been defined as a "strategic landscape approach to open space conservation, whereby local communities, landowners and organizations work together to identify, design and conserve their local land network, in order to maintain healthy ecological functioning."[1] As such, it is an organizing construct that fosters thinking of natural resources as a critical part of a community's life-support system. They are considered green because they are part of the natural environment, and they are considered infrastructure because they provide basic services people need for healthy living, such as filtering air and water or providing options for outdoor recreation.

Beyond human needs, conserving a biologically diverse landscape has intrinsic value for its own sake. Considering natural resources as GI, therefore, means that healthy, well-functioning ecosystems are important for the planet, even in areas where people never tread.

Even areas that humans may not directly utilize still have value. Credit: Green Infrastructure Center Inc.

GI planning evaluates various types of natural and cultural resources and prioritizes those assets that are most important and best meet current and future needs for a community or region. Thus, a GI strategy includes the process of identifying, evaluating, and prioritizing those areas deemed critical to preserving a healthy community. What follows is a summary of the six-step process recommended by the Green Infrastructure Center (GIC) and Esri. These steps will be covered in greater detail in chapters 4 through 8.

The steps

By considering natural resources (forests, wetlands, grasslands, beach dunes, among other natural elements), GI helps communities recognize the life-sustaining functions (often called ecosystems services) these ecosystems provide, as well as their tangible economic and social benefits. It also emphasizes that these natural resources need to be considered as a *network* because they are interdependent. A connected network of habitats allows species to migrate, fostering genetic diversity and repopulating areas that have been disturbed by a drought, forest fire, disease, landslide, or hurricane.

To create a GI plan, follow these six steps:
- **Step 1:** Set goals
- **Step 2:** Review data
- **Step 3:** Make asset maps
- **Step 4:** Assess risks
- **Step 5:** Determine opportunities
- **Step 6:** Implement a plan

Step 1: Set goals

What does the community, agency, or organization value? During step 1, determine the most important natural assets and functions and try to rank them.

All GI planning efforts must start with the establishment of goals. A map should not be generated without a purpose in mind. Before mapping, a geographic information systems (GIS) analyst or planner must consider what the community or organization most values about its natural resources. For example, what attributes does it value? Consider the following:

- Biodiversity in general, supported by connections across the landscape?
- Forests that provide clean air, water filtration, wildlife habitat, or timber for wood products?

- Water supply and groundwater recharge?

- Healthy fisheries?

- Sustained agriculture?

- Nature-based recreation, such as hiking trails and recreation areas?

- Protection of landscape settings around historic sites, burial grounds, or battlefields?

- Key views and beautiful vistas that will encourage tourism?

Community input is a vital part of GI planning. Credit: Green Infrastructure Center Inc.

Within a community, you'll likely establish these goals after a series of public consultation exercises that involve planners and mapmakers. This process can include public meetings, questionnaires, comments to a website, and so on. You should take a dynamic approach in updating maps as new data come in to support goals.

Although modelers and scientists can determine key habitats according to established ecological criteria, they will not initially know what the community's values are unless they have already been widely discussed. For example, does a local government desire to protect vegetated land cover to sustain the quality of its future water supply, or does it wish to expand housing into those areas? Does the community value better roads more than it does preserving scenic vistas? Does it value lower taxes more than it does spending money on planting trees? What will future community meetings and involvement reveal about the community's values and priorities?

These uncertainties may cause local GIS modelers to include a variety of landscapes in their GI network, even though these landscapes are not ranked highly by them in terms of, say, species richness or core quality. They may also include nonecological layers in their map as well, such as sites of cultural interest and local fishing and hunting spots. Later, these can be reranked and reassessed and, if necessary, new layers or filters developed to account for these priorities. In the initial map-making process, therefore, greater flexibility and a wider range of possible options should be considered than would be desirable in the final map or report.

Taking this approach, planners can easily respond to unexpected community desires. For instance, adding a spatial filter to capture all cores or habitat fragments near a drinking-water reservoir or those cores upstream from a water intake will help to include those cores important for protecting drinking water, if either of those turns out to be a major concern.

Step 2: Review data

What is known, or needs to be understood, to map the values identified in step 1?

Once goals have been established, planners can assemble and review all the existing relevant data necessary to create a GI network map. Mapmakers should research existing studies and available data: What are their findings, and how do they inform the GI plan? Within the lower 48 US states, users can download the Esri GI model of habitat cores to create a starting basemap. For counties in Alaska or Hawaii, users should look for existing studies or maps of GI cores and corridors from which to start.

If using an existing model or data layer, users should determine whether the data are accurate. For example, planners and cartographers should consider whether a core has been lost since the time the model was run. New housing developments, roads, or mines may have affected cores and reduced or eliminated them. By checking and comparing the date of the land cover used for the model with the dates of new land-disturbing activities, users can determine what, if any, updates to the map are necessary to remove areas that have since been developed.

Users should determine what data are still needed to inform the basemap (other state, county, or local data). Examples include watershed studies detailing existing water quality, wildlife studies, locally delineated wetlands, ecological inventories, element occurrence data from a state's or county's natural heritage program, or other surveys of natural elements.

Step 3: Make asset maps

How do you map the habitat cores and corridors of highest value and create a basemap of the GI network, as well as themed overlay maps based on the goals established in step 1 and data from step 2?

Once all existing data have been assembled, including any newly collected data to address unexpected goals, maps can be developed. In most instances, there will be a basemap of the GI network supplemented by themed overlay maps. The basemap is not a map of all environmental resources; it includes only those ranked as most important because they encompass an intact habitat core network that supports wildlife or because they fulfill a key goal, such as providing a buffer to a scenic resource or access to a park.

Depending on the project's goals and what the community values, the GI base network map should likely include the following attributes:

- Large, intact habitat cores
- Connecting corridors
- Major water features such as rivers and wetlands
- Background land cover and topography

For basemap design tips, see chapter 5, "Making asset maps."

Suggested examples of themed map overlays

Themed overlays can be used to show key relationships in how the base network supports related values and can include the following examples:

- **Water:** Watershed (catchment) boundaries, major rivers and key streams, areas that buffer municipal water supplies, surface water intakes, community wells, and groundwater recharge areas (if known), floodplains, and floodways
- **Working landscapes:** Agricultural soils, agricultural districts, or active forestry tracts; also include locations and routes for agritourism, such as fruit orchards and farms, wineries, honey producers, local livestock farms, and permanent farm vegetable stands
- **Cultural sites:** Battlefields and historic landscapes, churches or shrines, burial areas such as cemeteries or burial mounds, landscape-dependent tourism sites, scenic vistas, and scenic roads
- **Nature-based recreational areas:** Fishing, boating, hiking, biking, and birding; in urban areas, street trees, tree canopy, parks, and local streams

Step 4: Assess risks

What GI assets are most at risk and what could be lost if no actions are taken?

Once GI asset maps have been created, they should be evaluated to determine whether any areas or features are at risk. There, risks are easily discovered, such as upcoming proposals or plans to develop natural or agricultural lands or rezoning an existing farm for future industrial uses. Map developers should explore the extent to which current zoning adequately protects or threatens the county or regional land assets. There are also predictors of growth and development, such as proximity to roads or an abundance of small, not yet developed parcels. These risks can be overlaid in GIS to show areas at greatest risk. This process for conducting risk assessment using GIS is detailed in chapter 6.

Assessing risks to habitat cores and corridors

The following questions should be asked when assessing risks to habitat cores and corridors:

- Which areas are zoned for development, and do they overlap key cores or corridors?
- Where are new roads planned—will they fragment key cores or corridors?
- Which steams are impaired and need restoration; which streams are in good condition but may decline in the future (e.g., because of development pressures, land clearing, new industry)?
- What viewsheds or settings that support historical or cultural values are threatened by new buildings, roads, cell towers, or mines?
- Are there planned oil or gas pipelines that may disturb the area by fragmenting a core or corridor?
- Are there hazard areas, such as lands prone to landslides or flooding during storms?

Land conversion from forests and wetlands to subdivisions is the leading cause of habitat loss in the US. Credit: Green Infrastructure Center Inc.

Step 5: Determine opportunities

What is the potential for landscape protection or restoration?

Based on those assets and risks identified, the next step determines which assets could or should be protected, restored, or improved and which need the most attention as soon as possible.

Determining opportunities

The following questions should be asked when determining opportunities:

- Where should towns or developments be located in the future to allow for the retention of key resources, such as clean water supply or important forestry tracts or to take advantage of access to outdoor recreation?
- Where are new roads or transportation projects likely to affect corridors or cores—should those projects be modified to minimize or prevent impacts?
- Are there impaired areas where habitat can be restored?
- Does the zoning need to be changed? Does the area lack necessary zoning?
- Can a connecting corridor be replanted or widened with vegetation to provide greater habitat connectivity?
- Can we identify where we need to add key stepping stones to our habitat network?

Step 6: Implement a plan

How do you include natural asset maps in both daily and long-range planning?

Based on which areas and habitats have been valued as important in step 3 and which are at risk in step 4, an implementation plan should be created to protect or grow the GI network.

Creating an implementation plan

The following questions should be asked when creating an implementation plan:

- Should you update the zoning ordinances to better conserve high-priority assets and channel growth to appropriate areas?
- Can you acquire funding for habitat projects from government, foundations, companies, or individuals?
- How can you better educate landowners to encourage voluntary conservation action, such as donation of conservation easements?

- Is it possible to use the area's natural assets in marketing campaigns to expand tourism or attract new or compatible businesses?

- Can you site new public parks on areas that are highly ranked to ensure their protection (but not their overuse)?

- Where should you replant and restore native habitats and corridors?

- Do you need to create a schedule for updating the plan and host regular meetings to keep everyone engaged?

Areas can be restored to create habitats and reconnect the landscape.
Credit: Green Infrastructure Center Inc.

These steps are detailed in later chapters, along with tips for what data to utilize, how to work with the Esri GI model, and how to create a plan for implementation.

Choices we make for conservation

By Ryan M. Perkl, Esri's green infrastructure lead, who also serves as an industry practice lead for environmental service companies and for natural systems geodesign in the Professional Services division. He is a senior consultant and project manager at Esri.

Determining what to conserve is a ubiquitous challenge faced by virtually every individual and profession linked to the management of our air, water, and terrestrial resources. By extension, one might argue that the question is so fundamental that it is tied to our very existence as a species. It is within this context that modeling the natural landscape is a charge that must be undertaken with great fervor, intensity, and haste.

We are taught that our planet provides the building blocks necessary for life, that these ingredients are finite, and that they are perpetuated by a delicately balanced system of interrelationships. These concepts represent the foundation to build upon that we aim to model—what are our landscape's building blocks and how are they best arranged to ensure that these fundamental truths are best maintained?

We understand that these concepts are geographically rooted and distributed in both time and space. Moreover, we understand that there are interdependent processes at play operating across all scales, irrespective of jurisdictionally imposed boundaries. Admittedly, although portions of this understanding remain incomplete, it is nonetheless actionable and should serve as a guiding principle of our landscape modeling efforts—we must look at our landscapes based on the sum of their parts and place an emphasis on the interactions among those parts.

Much of what we love is local. The cold, clear water in the lake where we learned to swim, the trees in the forest where we first climbed, the view from our favorite lookout, the smell of the air in the desert after a monsoon rain, a sunset in the wilderness. These things, their places, and the crucial processes and patterns that link them together provide our motivation—to ensure that others may come to experience, value, and love such treasures in the same way we have, not to mention Earth's other inhabitants.

What we choose to conserve is driven by the former, and, in many cases, it is all that is lacking. More commonly than not, we have the theories, frameworks, models, data, and tools to achieve our goals. The way forward emerges from a foundation of universally relevant patterns and principles, is supported by systematically rigorous methods, and employs empirically driven and evidence-based solutions. Such efforts yield conservation networks instead of independent reserves, foster the inclusion of comprehensive assemblages of resources instead of narrowly focused targets, transcend multiple scales to capture system-level processes instead of focusing on compartmentalized boundaries, and recognize that what happens upstream has downstream consequences and that the impacts of our decisions do not occur in isolation—achieving these adjustments is progress.

This is fundamentally a spatial process that starts with the identification and mapping of the region's assets. By leveraging information such as this as part of our landscape models, we're better equipped to generate informed plans and make better decisions about the future of where we live. Through landscape modeling, we can leverage what we've been taught, deepen our understanding, nurture what we love, and apply it all in what we choose to conserve.

Step 1: Setting spatial goals and determining what should be mapped

Most successful projects have an overarching vision. If yours does not, then prior to developing goals, strongly consider helping the community develop a vision of what it would like its landscape to look like and how it should function. If a clear vision statement is in place, or if a comprehensive plan includes a proposed future, your GI maps may simply need to reaffirm and apply that vision to the particular process under way. Conversely, the GI maps may help implement a vision that lacks enough data to support it. However, it might still be important to ensure that the community understands the inspiration for the GI map and participates in translating it into specific planning goals.

Indeed, setting goals for a project is perhaps the most important step in the GI planning process because it forms the basis for what is mapped and how those maps are used to inform strategies. You may have heard that GI planning begins with making a map. Unfortunately, this rarely produces a useful map, because such a map must be informed by goals that prioritize what to include. For example, is the purpose to create healthier wildlife populations by connecting them to key habitats through corridors? Is the purpose to protect drinking water supplies and thereby select those cores that protect a community's water sources? Is the goal to provide more access for city residents to surrounding green spaces and thereby also select cores closest to population centers?

GI projects usually have more than one goal. Some goals may be primary and some secondary. A primary goal might be to create a connected and protected network of habitats to support native species or promote the economic revival of an area through ecotourism. A secondary goal might be to buffer local parks by protecting an adjacent natural area or replant a forest at an important historic site. In other words, because it's nearly impossible to protect everything, you should learn which cores are most important to select for the final GI network based on the goals determined at the beginning.

One way to represent that data is to choose all habitat cores based on size and intactness. In the US, this might simply be a map derived from Esri's Habitat Cores model. However, because it's difficult to protect everything from impacts (future roads must go somewhere, new families must live somewhere, and an area may be mined for coal or traversed by a power line at some point in the future), determining which habitats and landscapes are most important and why is a key first step. Preservation may be more important than development. Identifying areas deemed to have high importance will help protect them against such future possibilities and potentially require the relocation of a pipeline, road, or proposed development.

Mapping is an iterative process that involves local knowledge

Goals may be refined or expanded after mapping has begun. GI planning is an iterative process. Maps may reveal new ideas and highlight new priorities that were not initially considered. For example, once cores are downloaded and other data, such as the locations of parks, are overlaid, it may become apparent that some of the largest cores would be ideal to protect as parkland, especially if the maps show a lack of parks in an area. Similarly, maps may show that it is possible to create a new greenway by linking existing cores or by restoring degraded stream buffers to create a linked ribbon of green. Overlaying scenic views may identify new areas it is a priority to conserve.

Sample goals that cover some of these other ideas are shown in the table, "Sample goals for a GI plan."

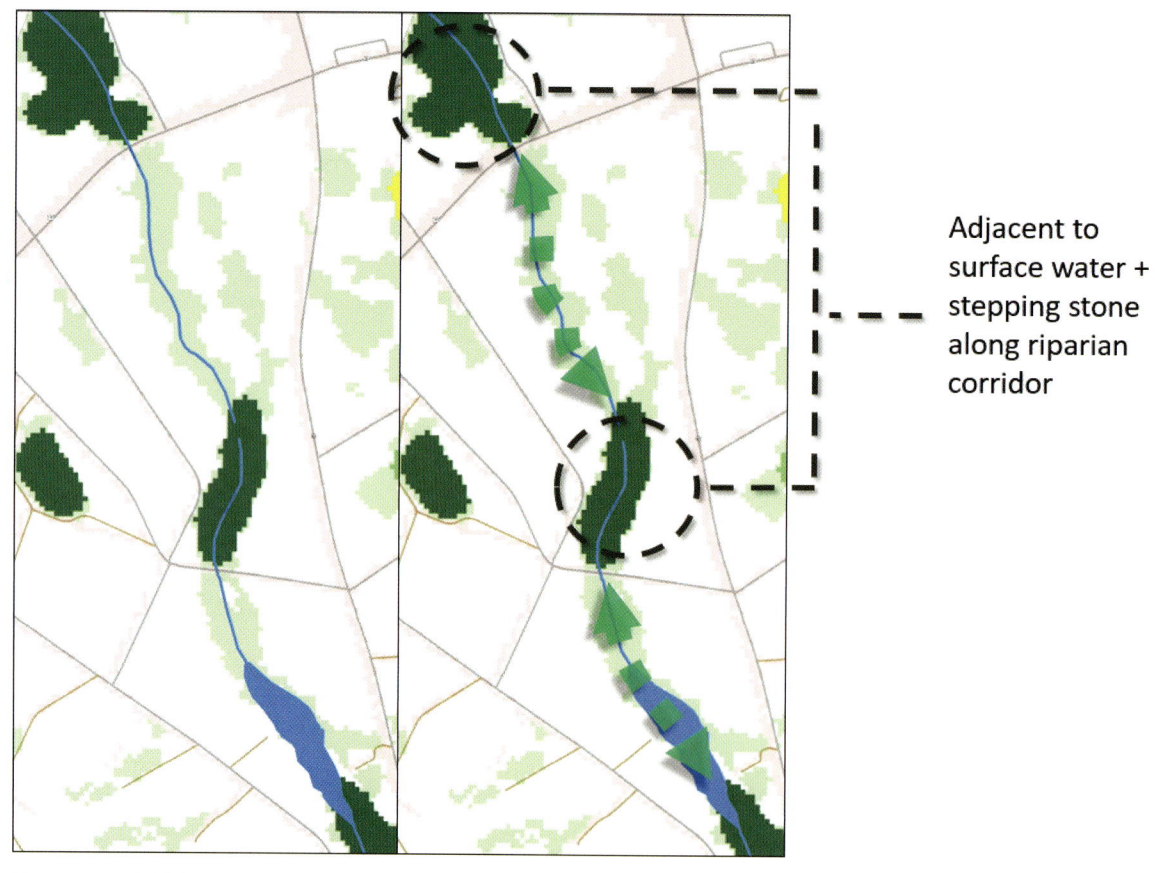

Adjacent to surface water + stepping stone along riparian corridor

Use maps to highlight opportunities. In this example, the need to restore the connection between isolated cores along a river is seen in the before (current condition) image (left) and after (adding a restoration strategy) image (right), in which a connecting habitat corridor has been created. Credit: Green Infrastructure Center Inc.

The process of creating maps allows new priorities to emerge. Maps may reveal that an asset thought to be abundant is in short supply, thus driving a new goal for restoration. Overlaying additional data layers such as historic or archeological sites may highlight previously unrecognized landscape features worthy of protection. For example, a forest may gain greater local significance because a historic event occurred there, such as a battle, or it had been an aboriginal settlement or a landing point for early settlers. In one case, a school in Virginia was located adjacent to a small woodland that had been a Civil War encampment. The proximity of the school to the wooded site added additional value to that site because of its ease of access for historical interpretation and science education. As a result, the site was deemed important to include. Without its historical significance, the woodland would have likely been developed.

The value of local knowledge

As these examples demonstrate, a site could be ranked more highly based on local knowledge of its ecological function or cultural significance. For example, a local river or wetland could contain a unique feature such as a heron rookery—something the mappers were unaware of— and cause the site to be more highly valued. Some of this local information may be found in a government's natural heritage data, but much of it will depend on local knowledge.

Hazard mitigation

Goals may also focus on creating safer, more resilient communities by avoiding a variety of hazards. Hazard mitigation is another planning need that is often mandated and can be met by identifying areas that are subject to flooding, landslides, wildfires, and so on. These areas may be set aside as places to conserve or avoid developing to prevent property damage and loss of life. They may also meet other goals for conservation. If a community is in a coastal or tidal area, a rise in sea level will probably impact habitats.

A flood-prone area is a likely candidate for conservation because it is unsafe for built structures. Conversely, if those habitats will be lost in 10 years' time, other habitat cores will become even more significant because there will be fewer of them. Similarly, an area subject to landslides might be included in a core network because it is unsuitable for development or recreation. Risk avoidance is discussed in greater detail in chapter 6.

Keep in mind the central purpose

Remember that a GI map has a central purpose to show us the location and extent of those green assets a community most highly values and that will most likely achieve its goals. The map also should create a more informed strategy for conservation or restoration, as well as for future comprehensive plans and zoning requirements. Consider asking this question for each issue, whether known or suspected: Can our green infrastructure strategy address the problem?

The Esri GI model offers a key advantage for this work. The model has already been created to select and rank habitats that are intact (have minimal fragmentation) and of a size beneficial for many species. The model also ranked habitats for factors such as water richness, landform diversity, and biodiversity. As chapter 2 explained, ample research has pinpointed factors that contribute to ecological integrity.

Although habitat protection to create healthy landscapes and ecosystems for wildlife (and people) is the key underpinning for this work, a GI landscape plan can help you meet many more goals.

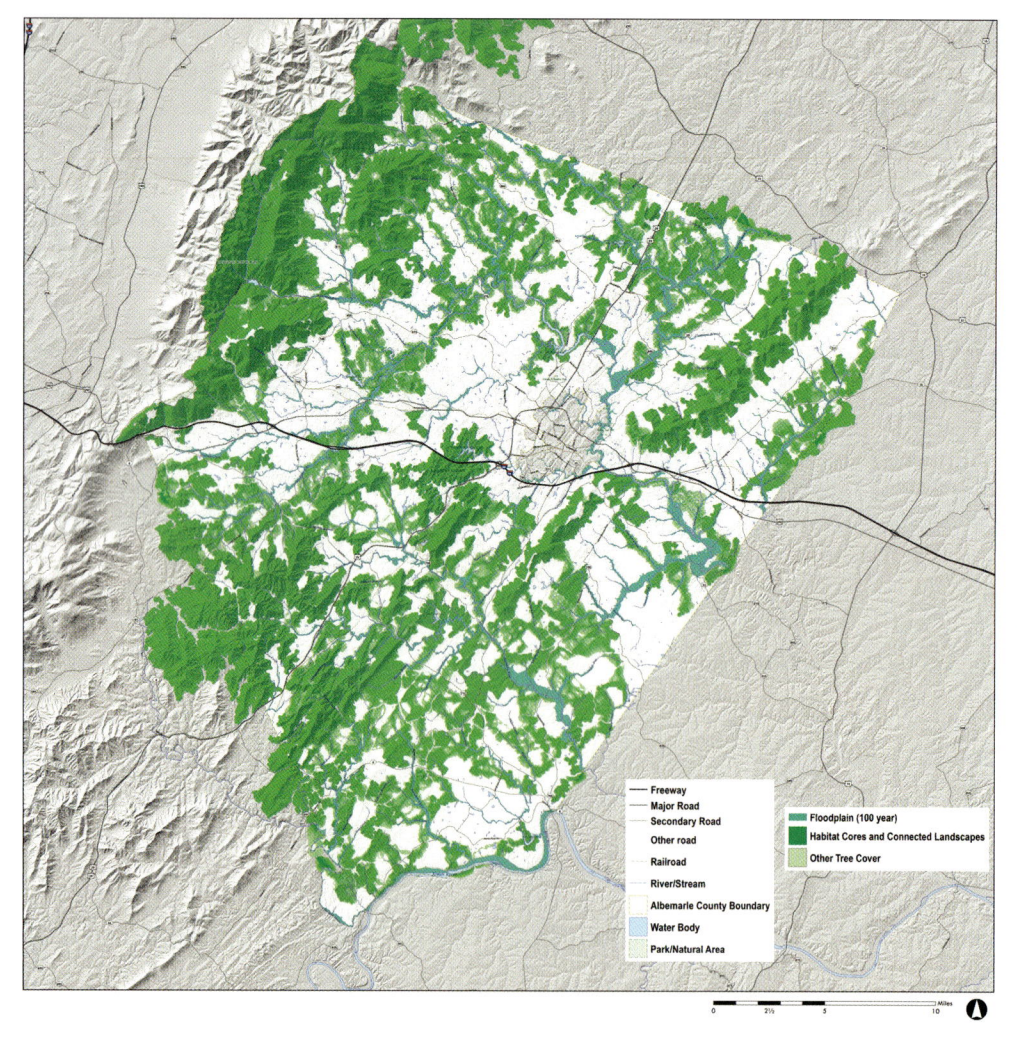

This map shows the base habitat cores for high-quality habitat cores in Albermarle County, Virginia. The basemap is currently used to inform strategies. Credit: Green Infrastructure Center Inc.

The system of rankings and prioritization helps planners select cores that support the community values. It also helps identify key corridors that connect highly ranked cores.

An overarching goal for a GI network might be to protect and connect habitat cores to support native species and create a healthful (resilient) ecosystem.

Or, if providing access to nature is a priority, an example goal for outdoor recreation could be to provide access to nature to support mental and physical well-being.

If protecting scenic resources is important, work must first be done to locate and evaluate those resources. One goal might be to safeguard the community's scenic and cultural heritage by protecting landscapes that support historic settings, pathways, and settlements.

If water quality and water supply are important for people, wildlife, and fisheries (and they usually are), a goal for watershed health might be important, such as protecting the quality of streams, lakes, wetlands, and bays. Another goal might be to protect key areas for groundwater recharge to supply a community's ground- and surface waters.

The following table includes example goals that you can craft to inform the creation of a GI network and what type of data might be used to address a goal. You can adapt any of these goals to best meet local needs.

Sample goals for a GI plan

Goal	Data Type to Meet Goal	Definition/Application	Source
Protect habitat for native species.	Intact forests or other habitat types (i.e., wetlands, intact grasslands, marshes, large dune systems).	Habitats that have adequate interior area which are mostly unfragmented and of a minimum size to accommodate a diversity of native animals, bird and plants.	Esri National Cores Map States may also have a habitat map
Protect a particular species (red cockaded woodpecker, big horn sheep, mountain lion).	Depends on habitat needs of particular species of concern.	Overlay the habitat needs of the species of concern with the cores map. If there are areas outside of the cores (e.g., a key water source or area used for breeding), add that to the map. The goal of the project may be to protect a particular set of species so multiple overlays may be needed.	Consult with wildlife biologists to determine likely habitat needs. Note that the US cores map of intact habitats will likely cover much of the area needed for protection.

Goal	Data Type to Meet Goal	Definition/Application	Source
Provide access to outdoor (nature-based) recreation to create healthy communities.	Location of existing parks (federal, state, local) and trails and greenways. Census data to learn where people are relative to parks. Include regional trails, rail trails, wildlife viewing areas, wildlife management areas, birding trails, etc.	Parks whose primary or majority of uses require natural areas. Overlay existing parks with habitat cores. Cores may fall outside of parks, within them, or may cross park boundaries. A core that intersects a park may become more important to protect since part of it outside the boundary is unprotected. Select areas that are close to existing or proposed trails, to either buffer the users' experience or provide for potential new connections in the future. Overlay census data with cores basemap and existing parks. Are there places where people lack access to parks? Are some cores more important because they provide views or access to nature? Conversely, are some cores more at risk because they are near developing areas (see chapter 6 for more on assessing risks to cores)?	State, regional, and local parks agencies will have data for park locations (or from the local government). Also include other nature reserves that are open to the public but privately run. For the US, US Census Data. May also consider racial and income disparities – for example, are parks distributed evenly without regard to race and income? If not, cores could be prioritized based on proximity to underserved communities.
Protect habitat. Protect water quality. Protect aesthetics. Support fisheries/fish nurseries (if tied to waterways or ocean).	Wetlands	Wetland types include forests, meadows, bogs, shrub swamps, ponds, lakes, streams or bays, and depending on location, may be tidal or non-tidal. Many species can only thrive in wetlands, and they provide nurseries for many birds, fish, crustaceans, insects, and animals.	In the US use the National Wetlands Inventory Data (NWI). The NWI may not be very precise. If local or county wetland data are available, add that to this layer. Delete incorrect data such as irrigation or stormwater ponds that may not be wetlands.
Promote agriculture row crops for a sustained food supply and rural economy.	Agricultural soils	Prime (best) agricultural soils occur in certain locations. If crops are important to the area, then agricultural soils can be mapped. Use land cover to select and remove areas already covered by urban uses (cities, towns, industrial parks) since not suited to large-scale farming.	In the US, USDA Soils Data Mart, select classes IV and V (top ranked).
Promote fruit orchards or vineyards.	Slopes Soil type Elevation	Fruit trees and vineyards do best on south or west-facing slopes in well-drained soils. Some fruits grow well only at higher elevations. A local extension agent can help suggest the best areas for orchards or vineyards.	Use a digital elevation model to select slopes. Use the USDA Soils Data Mart, select appropriate soil classes.

Goal	Data Type to Meet Goal	Definition/Application	Source
Protect surface water quality (streams, lakes, or bays).	Watershed boundary Forest cover Stream buffers Municipal water supply watershed boundaries Water quality data	Streams should be included in most GI maps as they provide habitat and are often good corridors for wildlife, as well as sources of drinking water. To determine how well forested the watershed is, the forest cover can be clipped in GIS to match up to the watershed boundary and used to determine the percentage of area covered by forests or how much area of the watershed is covered by intact landscapes. For water quality, map stream buffers by using GIS to find center lines of streams and map 100-foot widths on either side to see extent of forested stream buffers for buffering runoff. For large rivers, use stream edge, if known. If using streams for wildlife corridors, select 300 meters on either side of stream and intersect with forest layer to see if adequate forest buffer to provide a protected corridor. If protecting headwater streams, use steep slopes and elevations to select upland streams for protection.	National hydrography data set for stream locations and augment with additional local data. See forest canopy above. In the US, 305B reports contains water quality ratings and the 303D lists contain waters found to be impaired and in need of restoration.
Protect cultural resources by preserving their landscape context.	Historic sites (in rural areas), battlefields, cemeteries, tribal lands, etc.	Historic sites are often dependent on the context of the surrounding landscape. Either buffer each point (building) by 300 meters, or sites can use outline or buffer them. You may also want to protect the views from this site for visitors.	Obtain historic data from State Division of Historic Resources. Some sensitive data, such as Indian burial sites, may not be available. Viewsheds can be mapped using the Create VIewshed Tool available with Spatial Analyst or the Viewshed Analysis Tool available with Esri's 3D Analyst license.
Prevent urban heat islands Protect aesthetics Reduce stormwater (developed areas) Sequester carbon to mitigate climate change Clean the air	Tree canopy	Canopy is the coverage by forests (bird's-eye view) and is more commonly applied to urban areas where other values (besides interior) also become important, such as cover to keep cities cooler, aesthetic values of trees to downtown areas, and habitat for urban birds and other animals. Trees also mitigate urban stormwater and sequester carbon and clean the air.	Forest canopy may be available from a state forestry agency. In urban areas, along with the canopy data, inventories of street trees or park trees are important for knowing the diversity and condition of the trees. Tree canopy data should be less than five years old to be useful since urban land cover changes often. In the US, I-Tree is a software tool to help determine benefits provided by the canopy.

Credit: Green Infrastructure Center Inc.

Ensuring the community understands and supports the project

When a GI project is initiated, the challenge is to create some consensus around a limited, defined set of the most important goals.

> **Examples of goals that a community might support**
>
> - Protect a rural economy (e.g., forestry, farms, and grazing lands) from development.
> - Beautify entrance corridors and the downtown area to encourage tourism and economic development.
> - Preserve regional forests for wildlife, hunting, and leisure opportunities.
> - Ensure biodiversity and a healthy ecosystem.
> - Recharge groundwater aquifers for drinking water by protecting forested land cover.
> - Conserve community character and heritage by protecting a historically important landscape.
> - Promote water-based recreation, such as kayaking, fishing, and boating.
> - Save money by directing development into areas where services (e.g., roads, schools, power lines) already exist and avoid costs of sprawl.
> - Reduce stresses on wastewater plants and local waterways through a comprehensive system of stormwater mitigation measures.
> - Create a resilient community by avoiding hazards and protecting the public from unstable slopes, floodways, fire-prone areas, or karst areas subject to sinkholes.

Stakeholders may not initially understand their roles in a GI plan. For example, local and regional historical societies may not see their role as working on landscape plans, even though the plans included maps of key cultural and historic sites. They may need more information to understand why a landscape plan is central to their goals for preserving historical character.

Once a community has identified its goals, it can create criteria to select cores by adding those priorities. In the example that follows, a community first selected all the highest-ranked habitats identified by its GI model (cores in the top two quintiles when ranked based on ecological integrity). They then added additional (lower-ranked) cores to their network, if they met one of the following criteria:

- Cores that touch, intersect with, or are within 500 feet of a water body
- Cores that contain or are within view of a cultural or historic site

- Cores alongside identified scenic roads
- Cores that form a link (corridor) between two higher-ranked cores
- Cores near a park or other public land
- Cores that touch (as a buffer) or contain a conservation easement (or other government-protected land)
- Cores that provide access to recreation or that buffer a recreation area
- Cores that provide a link between the town and surrounding area
- Cores that contain a unique geological feature (a mountain bald, a Carolina bay, or a cave)
- Cores that contain a rare, threatened, or endangered species
- Cores that contain a hazardous area, such as a floodway or landslide zone

In prioritizing the GI network, remember the community's stated goals. If the map is for a land trust or other conservation group, does the selection of key priorities meet the group's mission? Ensure that you create enough mapping rules (guidance) on what to include on a final GI network map. Following is an example map from Darlington County, South Carolina, showing cores before and after additional prioritization criteria were added.

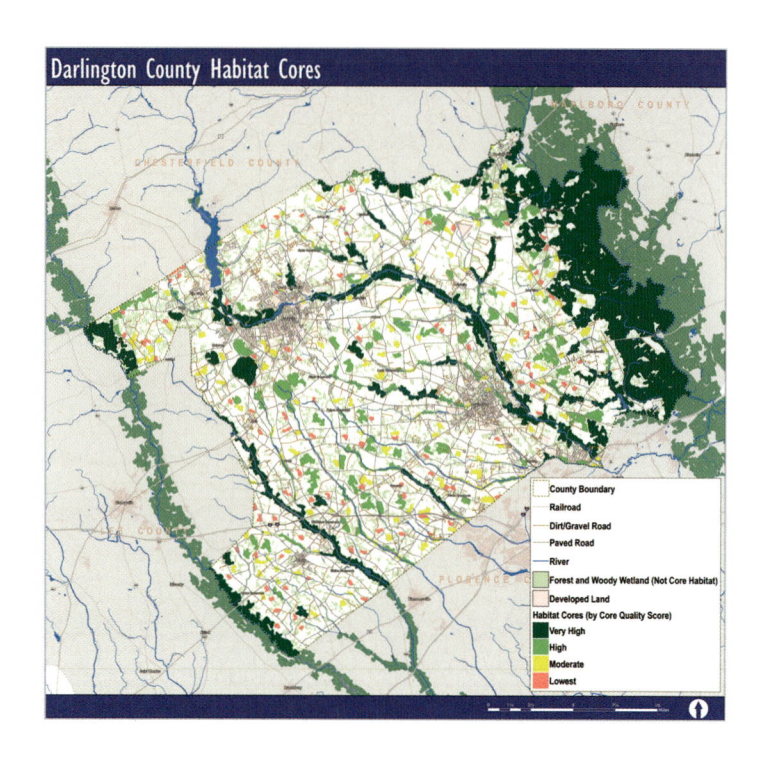

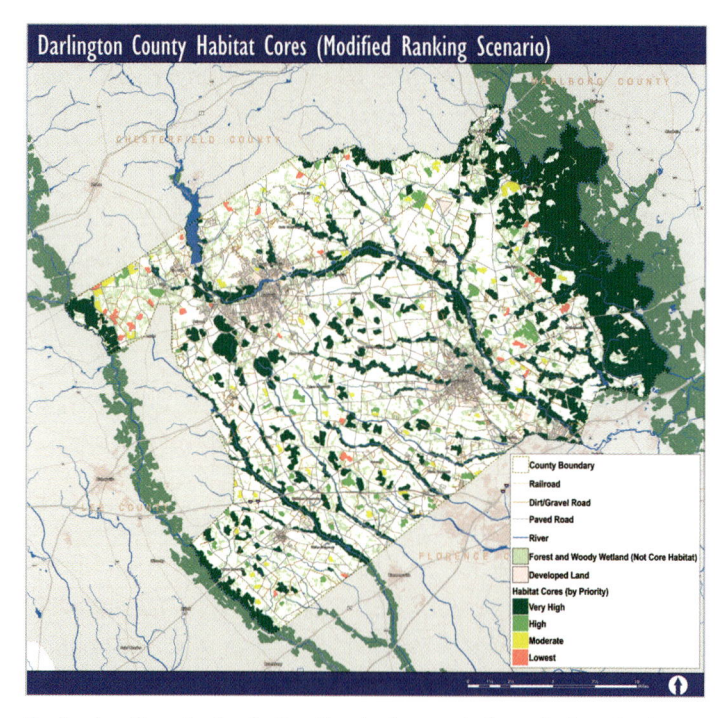

Darlington County, South Carolina, before and after priority criteria were added. Several cores have increased in rank—see color change—such as values for rare habitats, buffering high-value cores, and access to natural areas. Credit: Green Infrastructure Center Inc.

Green infrastructure: Map and plan the natural world with GIS

Cores may also be selected because they provide habitat in an area that would otherwise not have any habitats if only the highest-ranked areas were provided. For example, cores may be selected for an area, even though they are ranked lower; otherwise, those areas would not have any habitat to protect.

Time frame and specificity for goals

Perhaps the most important consideration is that mappers should translate all goals into a spatial format—the goals should be mapped. Are the adopted goals specific enough to provide direction to create a map or overlay? Goals such as "preserve all natural areas," "preserve all ocotillo," or "help the monarch butterfly" are not helpful for mappers. They need specifics that they can map and that make sense to map.

If not, you should modify your goals to add specific qualifying statements. For example, the goal "to keep the county's water clean" needs clarification, such as "by requiring forested buffers of at least 100 feet along both sides of all perennial streams." Or, if wildlife movement between two cores is important, the goal could include "to protect buffers of 300 feet minimum on each side of Stream X." You also may need to add specific objectives, such as details on whether there are areas of higher priority, such as cores that protect headwater streams or streams that feed into drinking-water sources. A panel or committee of topical experts can help elaborate on and suggest refinements to specific parameters. These experts should coordinate with the mappers to identify mapping issues and suggest refinements back to the community or its leaders.

As noted, GI planning is an iterative process. Once the data-gathering and mapping process has begun, you may need to update the goals to reflect newly identified priorities. Revisions can happen anywhere in the process. During the risk assessment phase, a new goal often suggests itself on the basis of what has been learned. For example, protecting scenic areas may become a priority once you determine that a new ski resort could threaten an iconic mountain view.

Using the Esri cores model

The authors recommend the methods of the Esri cores model for identifying landscape areas as habitat cores. As noted earlier, the model was created based on several decades of applied research into the type and extent of landscapes that are likely to support the most species. The model shows cores that have already been identified as highly ranked. Local users should then gather additional data to determine specific goals for development. That said, local experts may offer valuable guidance. For example, cores with a perennial stream may be much more significant in an arid landscape than one in a wetter climate where stream density is high. Vastly more species may live in the riparian area along a stream that flows through an arid landscape, as the riparian area provides the only suitable habitat for many species. Consultation with experts is thus prudent.

> **Decision metrics**
>
> Creating decision metrics is a way to craft mapping rules to help along the way. A decision metric is a standard that helps the designer prioritize what to conserve first and why. Creating decision metrics early in the process can help resolve potential conflicts later. These metrics define priorities based on a community ranking of what is most important and might include such things as "First, protect all cores that shelter rare or endangered species."

Gaining community support

Prioritizing natural assets for conservation is a value-driven process. Determining what is valuable requires community engagement. For example, although the best available science can show the types of habitats that are important for wildlife, the community should first decide that wildlife conservation is important. This process may require education about the values of wildlife prior to asking for input on goals.

Community members should join the GI planning process as early as possible, not after the plan is finished and the maps have been created. Local citizens need genuine roles in setting or reviewing a project's goals to ensure they have buy-in and are fully engaged. Their comments must be heard and taken seriously. An interactive website or regular meetings will help citizens stay updated on the progress and any issues. Finally, they should be reengaged once the plan nears completion while there is still time to offer meaningful input and sign off on the plan.

Stakeholder input should provide opportunities for education and for new community insights. Credit: Green Infrastructure Center Inc.

Without stakeholder involvement, local goals for the GI network that could help refine the outputs will be lost. Goals may meet larger aims, such as the European Commission's biodiversity and ecosystem service goals mentioned in chapter 1. But global action starts locally. Academic or purely government-led endeavors can fail to capture specific local goals and local enthusiasm. Without community involvement, local habitat cores that support historically and culturally significant sites may be overlooked. Top-down projects may miss opportunities to include scientific knowledge about the location of local species and habitat threats, such as a farmer illegally draining a wetland used as a local dumping site.

Public support is vital for implementing a GI plan

Finally, building support for conservation goals requires early stakeholder engagement to ensure that more specific goals are crafted and implemented. Most models fall short at this stage—all maps must be implemented by a wider audience than the academicians or government officials who created them.

Simply put, communities are more likely to support ideas that meet goals they helped to create. For a detailed description of how to engage the community, see the author's prior books on strategic GI planning.[1]

The work of a committee of experts

You may need to create a committee of experts to build support for goals from stakeholders and technical experts. Chapter 5 discusses in detail the topic of using experts, but note here a word of caution. Many times, experts recommend extensive field studies as the first step. In one project facilitated by the GIC, a local expert recommended hiring 500 biologists to survey the county, even though there was no funding for even one biologist.

The perfect data trap

Although more data are useful, planners and analysts should avoid the perfect data trap. A GI model is intended to show areas that *most likely* support an abundance and diversity of species or *most likely* support identified goals. You should collect extra data from field surveys only as time and funds allow. And that data must be complete, or they will give an imbalanced assessment. The key is not to visit only one place and declare it the best simply because it was surveyed. For example, North Carolina created one GI model that included only areas that had already been surveyed, which meant that those areas not surveyed were deemed as having no ecological value simply because there were no data on them.

Models are predictive tools. When gathering data to ground-truth a model, it's important that the survey is designed to sample the habitat network consistently. One community group collected a great deal of field data from locations where they gained access for surveys. But this expedient approach did not represent the full extent of the forest's biodiversity. In fact, what were thought to be rare ecosystems because only a few were initially discovered turned out to be common—the group simply had not obtained representative samples for the area.

Concluding thoughts on step 1

Completing the first step will result in more than one map. You should have a basemap of the habitat network and several themed overlays, each highlighting a particular issue or resource. Trying to fit everything on one overlay would create an unreadable jumble. Selecting focus themes allows the map team to highlight certain key features of interest and separate them into specific topics. For example, a themed overlay map that depicts forested lands can be correlated with one that shows agricultural soils to identify suitable areas for new forestry tracts or show where existing forestry would be better utilized as farmland. A themed overlay of cultural resources or of tourism sites can show how key cores and corridors support those uses.

Themed maps can also further prioritize the core network map. For example, does a core also support fishing or hunting? Or does it have a hazard area within it, such as a floodway or landslide-prone area, meaning it should not be developed and should be added to the network of protected lands? Does the core or corridor protect a vital drinking-water supply? And so on.

Green infrastructure can also protect drinking water, such as this reservoir for New York City located in Ulster County, New York, surrounded by forests. Credit: Green Infrastructure Center Inc.

Having a clear purpose in mind, even though you will tweak it along the way, will help keep the process on track. It will also create a more expedient process in the long run. Again, ensure that you discuss and discern the key reasons for a project with a wider audience and stay open to new ideas.

It is also critical that the mapping process remain focused. Although GI plans can meet many needs, mappers should ensure that the purpose is distinct enough to avoid too much confusion. Many projects become lost in endless map versions, but you can avoid this problem if you are clear about the reasons for basemaps and associated overlays.

Planners often make the mistake of not clearly or adequately communicating their goals to the GIS team. A successful GI project requires collaboration from planners, GIS analysts, park and open space staff, land trusts and other conservation groups, developers, stakeholders, and community members.

This chapter briefly covered the six steps in GI planning. The chapter reviewed the notion that GI planning is an iterative process and new goals may emerge from the data and maps. Finally, it explained why and how to build community support to ensure the plan can be implemented once it has been finalized. Chapter 4 discusses data acquisition in detail as part of a strategy to obtain the right data to create a map, beginning with the Esri cores model. Chapter 5 then demonstrates graphically how you can use data and prioritization to create local maps with lasting utility.

Note

1. Firehock, Karen. *Strategic Green Infrastructure Planning: A Multi-Scale Approach.* (Washington, DC: Island Press, 2015).

Chapter 4

Getting the right data

Chapter 3 discussed the need for setting goals as the first step when planning for conserving or restoring GI. These goals drive subsequent data collection and mapping efforts. Thus, the data required to map high-priority landscapes will vary, depending on the specific priorities of a community. In addition, every community has a unique set of assets that make it special. These include both natural and cultural assets.

Where to begin?

Which datasets are needed for GI planning? To answer that, it will be helpful to review *why* data are collected in the first place. There are three overarching reasons:

1. Data identifies the location of GI assets, which can fall under many categories (e.g., biological, cultural, recreational). These data are the primary content in a GI plan.

Example datasets that help identify GI assets

Data	Purpose	Source
Water Resources		
Watershed Boundaries and Major Streams	To manage by watershed and also to determine potential runoff issues.	National Hydrography Dataset.
Healthy Streams	It is important to flag streams of exceptional quality. These can be included on a map of best water resources. Also consider adding naturally reproducing trout waters (not stocked but self-sustaining) and waters with exceptional or rare species.	Environmental Protection Agency, Department of Fish and Game, Department of Environmental Quality, or similar.
Wetlands	Provides sensitive landscape and key hydrology.	National Wetlands Inventory (note this is not very precise, especially for small and forested wetlands). County/City/Local GIS.
Wetland Banks	Identify protected wetlands.	Municipal GIS

Data	Purpose	Source
Fisheries	Includes known natural trout waters and streams that support other key species.	Department of Fish and Game, or similar.
Public Wells	These serve 20 or more people. Correlate wells to land cover to determine the level of protection. Public wells within corridors and cores provide additional justification for conservation prioritization.	Municipal GIS, Department of Health, Department of Environmental Quality, or similar.
Public Reservoirs	Areas draining into reservoirs should be as forested as possible. A forested buffer around a reservoir is also important. Although water will be treated, treatment costs increase if water is not clean to begin with.	Municipal GIS should also include drainage area.
Groundwater Recharge Areas/ Aquifers	Protecting recharge areas for drinking water is important. Areas with known karst also can be mapped since there will likely be a much more direct input of surface water into aquifers.	Usually not available unless a study has been conducted.
LID/BMP Features	Best Management Practices for low-impact development (rain gardens, green rooftops, pervious pavement, rain barrels, etc.).	Municipal GIS.
Recreation		
County/City Trails (future and existing)	Determine trails that intersect and utilize natural assets and areas of natural assets that support views from trails. Include water trails, if any.	Municipal GIS.
Federal Trails	Determine trails that intersect and utilize natural assets and areas of natural assets that support views from trails. Federal ownership denotes a higher level of protection from change.	Various federal agencies and state GIS.
Other Regional Trails	Same as above, and may show areas where interjurisdictional cooperation is needed.	Varies.
Hunting Lands	Almost always privately owned and leased to hunt clubs. Leases are usually for set periods of time. May show preferred land use and importance of ensuring a connected network of land.	Can be difficult to find one data source. May need to ground truth or contact hunt clubs individually to learn which tracts are leased for hunting.
Wildlife Management Areas (WMA)	Shows level of protection and use. Pay particular attention to lands outside areas that may also need protection since they are a magnet for adjacent development and adjacent land uses may impair WMA.	Department of Fish and Game, or similar.
Birding and Wildlife Trails	Often privately owned and not formally protected. May provide greater priority for conservation of land underneath or adjacent to it.	Department of Fish and Game, or similar.
Boat Ramps and Launches	Denotes public access area that may need protection. Consider different symbols on map for motorboat launches versus canoe only.	Department of Fish and Game, or similar.
Boat/Kayak Paddling Trails	Shows routes for non-motorized recreational paddling.	Department of Conservation/ Recreation. Department of Fish and Game, or similar.

Data	Purpose	Source
Scenic Rivers	Denotes areas that may offer special recreation and views. Also consider this for the water map and heritage map.	Department of Natural Resources, or similar.
Historic and Cultural Assets		
Historic Register Sites	Particularly sites in rural areas influenced by landscape setting. May also include suburban areas. Consider placing a buffer around these. How do the mapped natural assets support these historic sites?	Municipal GIS (federal, state and local) and State Division of Historic Resources.
Potentially Eligible Historic Sites	Same as above; also protection of adjacent land may protect these sites.	Municipal GIS.
Historic Districts and Rural Historic Districts	You may want to include all districts. What are significant for natural asset maps are those districts supported by natural landscape features (viewsheds and buffering adjacent land uses).	Municipal GIS and State Division of Historic Resources.
Battlefield Areas (National Register and Eligible)	You may want to maintain sites for historic reasons and determine whether, and how, natural assets include and buffer these sites.	Municipal or state GIS.
Scenic Roads (byways)	Overlay with natural assets and determine whether, and how, assets support viewsheds of these routes.	Department of Transportation.
Rural Historic Districts	While not a protected landscape, designation shows historic significance. Consider prioritizing natural assets within the district.	Municipal GIS and State Division of Historic Resources.
Century Farms	These are farms at least 100 years old. Based on nominations, not comprehensive. May also include on working lands map.	Municipal GIS and State Agency for Historic Resources.
Working Lands (Agriculture & Forestry)		
Parcels >25≤100 Acres	Useful to determine areas less or more suitable for commercial farming or forestry.	Municipal GIS.
Parcels >100 Acres	Same as above; also consult with your forestry agent to determine the ideal minimum size for sustained forestry (e.g., 25 acres, 50 acres). Contact the extension service about farm size for fruit and row crops.	Municipal GIS.
Prime Agricultural Soils	Useful to determine areas suitable for row crops. Overlay with zoning and development to take out soils with incompatible land uses.	Municipal and NRCS Data Gateway. Spatial data are available for most (but not all) counties. http://datagateway.nrcs.usda.gov/
Active Forestry Lands	Shows lands that are actively being managed for forestry, such as plantation forests. Can be used to distinguish different types of forest.	Department of Forestry.
Orchards and Vineyards	Identifies orchards, vineyards, or similar land uses that contribute to a region's economy.	Municipal GIS.
Active Farms	Identifies lands that contribute to the agricultural economy. This is difficult to define and accurately map, but some local datasets may be available.	The USDA National Agricultural Statistics Service produces the Cropland Data Layer, which can be used to estimate where crops are being grown.

Data	Purpose	Source
Wildlife		
Habitat Cores and Corridors Model	Habitat Cores and Corridors; i.e., interior forests, dunes, wetlands that provide wildlife habitat and connections. Check with your state or regional agencies to see if a model has been built. Otherwise, see chapter 7 for guidance on how to begin the process	Division of Natural Heritage, Department of Natural Resources, or similar.
Forest Cover/Tree Canopy	Use to show forest cover relative to habitat cores and corridors. Also can be used to determine forest cover for a watershed to consider runoff potential and impacts to water quality. Can often be derived from land cover data.	Various sources (Federal, State, Municipal). Select the most recent and highest resolution data. Check if an urban tree canopy (UTC) assessment has been conducted for your study area.
Essential Wildlife Habitat	Selected based on land cover and acreage.	Division of Natural Heritage, Department of Natural Resources, Department of Fish and Game, or similar.
Important Bird Areas	This is useful if data are spatially recorded and discreet. Flyways and known nesting areas can be protected or used to add reasons for protecting a particular region.	Typically too large to be meaningful for local planning, and these areas may move across the landscape (they are often not static).
Species (rare, threatened, endangered)	This data is not usually made available publicly. States often have a subscription service for counties to look up species by area.	Contact Division of Natural Heritage, or similar.
Rookeries or other unique habitat areas	Shows key habitat areas and useful to know, in order to protect. Data should be generalized (GIS buffer) to prevent the public from disturbing the area.	Local knowledge (location can change). Also contact Division of Natural Heritage, or similar.

Credit: Green Infrastructure Center Inc.

2. Data that assess the level of risk to GI assets: How likely are the mapped assets to disappear or remain the same? To answer this question, specific data must be collected, such as zoning areas, planned subdivisions, and conservation areas.

Example datasets that help identify risks to GI assets

Data	Purpose	Source
Zoning	To evaluate allowed land uses and potential risk or compatibility with priority habitat cores.	Municipal GIS.
Future Land Use	To evaluate future risk or compatibility.	Municipal GIS.
Conservation Easements (county, state, and nonprofit/private)	To determine what is protected and least likely to change. Overlay with priority habitats to determine their level of protection.	Municipal GIS, Land Trusts, State Division of Natural Heritage (confirm with muncipality data to ensure it is up-to-date).
Publicly owned land	To understand the likelihood of the land to change. Dependent on the locality's intention for the land (sell, develop, protect, remain in current use, etc.)	Municipal GIS.

Data	Purpose	Source
Impaired Streams	Useful in determining risk and where additional forest cover or stream buffers needed. Also to evaluate risk of new impairments that may occur.	Municipal GIS, Department of Environmental Quality, or similar.
Floodplains and Floodway Fringe	To determine areas of risk that may be best left undeveloped for public safety while also providing wildlife corridors. Overlay with forest cover to determine buffer capacity.	Municipal GIS.
Steep slopes		
State Forests	Use to determine the level of protection for natural assets.	Department of Forestry.
Water Monitoring Stations	Can be used to obtain stream data and current water quality. Also use impaired waters list to determine level and source for impairment. Consider whether impairment type can be addressed by land conservation measures.	State Department of Environmental Quality, or similar. 305(b) Report for Water Quality. May also review the 303(d) list from the DEQ for impaired waters.
Storm Surge/ Sea Level Rise Projections	Used to assess risk and plan for resiliency in coastal areas. Typically, these data are the result of a model.	NOAA, universities, or marine/coastal groups.
Federal Parkland	Used to determine the levels of protection for natural assets and who are the management entities.	Municipal GIS.
State Parkland	Used to determine the levels of protection for natural assets and who are the management entities.	State or Municipal GIS.
County Parkland	Used to determine the levels of protection for natural assets and who are the management entities.	Municipal GIS.
Parcel Information	Parcel size and ownership are helpful for evaluating longterm conservation potentials (e.g., are they large enough to manage for habitat or working lands?). For urban areas, knowing where vacant parcels are located can help identify opportunities for restoration and creating new green space.	Municipal GIS. For working lands, to determine size, consider sortting data by ownership and adjacency, as farms and forests are often made up of several parcels under one owner.
Building Locations/ New Constructions	Used to determine exact locations of occupied dwellings to determine if any new fragmentation of cores has occurred.	Municipal GIS – E911 Point Data.
Farms with PDR Acquisition	Shows land with permanent protection from development. You may also use it to compare farms with nearby and adjacent incompatible land uses.	Municipal GIS.
Agricultural and Forestal District Lands	Shows land intended for agriculture and land with some temporary level of protection. Compare to adjacent and possibly incompatible land uses.	Municipal GIS.
Service District Boundaries	For orientation and management plans and to show areas most likely to develop.	Municipal GIS.

Credit: Green Infrastructure Center Inc.

3. Reference data that provide context for the maps: These data include base data, such as roads, governmental boundaries, place names, and major points of interest. When you make maps of features, it is critical to include reference features to help viewers orient themselves. This information is often already housed in local government GIS departments.

In traditional land planning, "gray infrastructure" systems, such as road and utility networks, are laid out with little regard for the character of the landscape. Landscape elements, when they are considered, are usually treated as constraints that must be planned around. For example, mountains and water features must be tunneled through, bridged, or piped when laying out a road network. Parks, trails, and conservation areas are often shoehorned into this scheme after development has occurred. In contrast, *GI planning considers the character of the landscape first*: What defining topographical, hydrographic, and biological systems are at play? Understanding the intrinsic character of the landscape is critical for maximizing the benefits provided by GI.

The first step is to map high-quality habitat cores and corridors. This map is referred to as the *green infrastructure basemap*. This basemap identifies intact habitat, connections across the landscape, and major landscape elements, such as topography and water features.

The purpose of the basemap, as the name implies, is to serve as a base that provides context as other GI assets are mapped and laid over it. Thus, once the basemap has been completed, other assets, such as prime farm soils, nature-based recreation sites, important viewsheds, and historic buildings, can all be mapped in context. This mapping easily shows relationships between various GI assets and is extremely helpful when identifying where the landscape supports other community values, such as a habitat core that contains a trail, supporting both biodiversity and recreation.

GIS analysts are familiar with the term *overlay*, as when layering datasets to explore patterns, and it is a central concept in GIS. Meanwhile, the basemap serves as an *underlay*, showing the backbone of the GI network consistently across all maps that are overlaid across it, referring them all to a single context.

> *The GI basemap identifies intact habitat cores, connections across the landscape, and major landscape elements, such as topography and water features. It provides a consistent context for other maps of GI assets.*

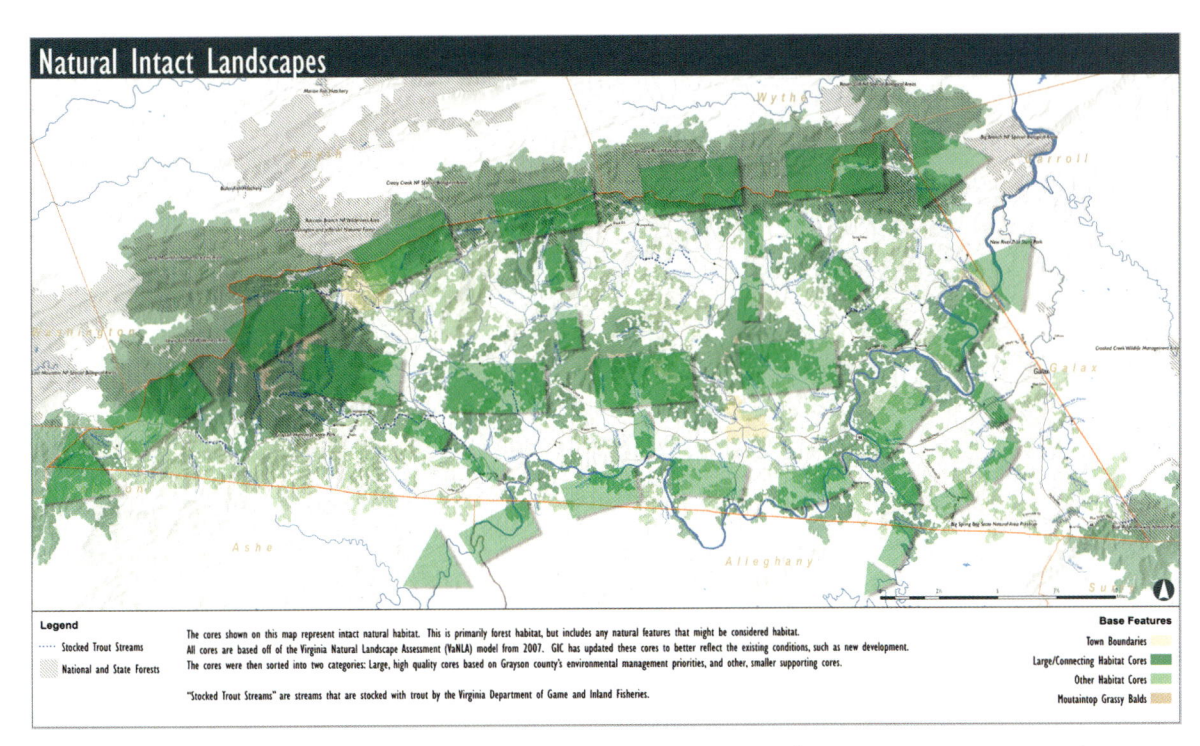

This map of Grayson County, Virginia, shows large, intact areas of habitat (cores) and general linkages across the landscape. Credit: Green Infrastructure Center Inc.

Collecting the data required to build a basemap can be challenging because "off-the-shelf" data are not always available. Collecting the data needed to make a roadmap, for instance, is usually quite simple, because roads data are widely available. However, landscape-scale habitat data are often varied, inconsistently collected, and not publicly available.

In addition, habitat data are collected and created for many purposes and are not always useful at a local scale. For example, when a black bear habitat is mapped, the resulting map covers swaths of territory. Therefore, if all land in a county is classified as bear habitat, as the author once discovered in a GI project, that data will not help a county prioritize habitat for bears within its boundaries.

Fortunately, landscape-scale habitat data are widely available, as discussed in chapter 2. GIS analysts and planners should check with state or county governmental conservation division or natural heritage staff to learn what data and tools are available.

Although several states have built models, by 2016, the US was still a long way from coast-to-coast coverage. Due to these widespread gaps in such data for the US, Esri partnered with the Green Infrastructure Center to construct the national model layer of intact habitat cores. This information provides complete coverage for the contiguous US and is an appropriate place to start when building a basemap of habitat cores.

Chapter 4 | Getting the right data

Intact habitat core data for the nation

You can access Esri's intact habitat cores data three ways to begin building a basemap:

- Download the data.

- Interact with the data using apps.

- Use an ArcGIS toolbox to identify cores manually.

The data

You can download the intact habitat cores layer and many supporting data layers for each of the lower 48 US states. This option is the simplest way to access data for your area of interest if you are comfortable using ArcGIS.

A gallery of map packages, one for each state in the contiguous US, provides easy access to the cores data and the data used to derive the cores and is available at **http://esriurl.com/GIData**.

The map packages include the following data:

- **Intact habitat cores:** Cores are intact habitat areas at least 100 acres in size and at least 200 meters wide. These areas are derived from the 2011 National Land Cover Database (NLCD 2011). Potential cores areas are selected from land cover categories not containing the word *developed*, or categories associated with agricultural uses (crop, hay, and pasture lands). The resulting areas are tested for size and width and converted into unique polygons. This process resulted in the generation of more than 550,000 intact habitat cores. These polygons are then overlaid with a diverse assortment of physiographic, biological, and hydrographic layers. Fifty-three metrics are calculated for each core, which can be used to guide conservation priorities (a full list of attributes and their descriptions is available in Appendix B).

- **Habitat fragments:** Any areas of natural land cover categories not containing the word *developed* (or categories associated with agriculture uses) that do not meet the minimum size and width criteria to be considered a core. Typically, these areas are not considered core habitat because they are smaller, narrower, and more fragmented. They can make up a large portion of the landscape's surface in areas significantly fragmented by roads, railroads, agriculture, and other human-dominated activities. Although smaller than a core, these areas may represent highly valuable local resources and serve as complementary assets that improve the ecosystem functions of neighboring cores, in habitat in highly fragmented landscapes void of cores, in stepping stones for facilitating local connectivity, or by providing opportunities for habitat restoration.

- **Connectors:** This dataset is composed of least-cost paths between habitat cores. These least-cost paths are the modeled paths of least resistance for plants (seeds) and animals to move across the landscape and can serve as a starting point for identifying lands that are important for maintaining or restoring as corridors in a GI network. The connectors are based on a *cost surface* that reflects the relative ease of movement for terrestrial species and consider several factors, including NLCD classes, slope, proximity to water, and habitat core score. Generally speaking, natural land cover classes and areas proximate to water are parametrized to exhibit lower costs to species movement, whereas developed areas and areas proximate to built infrastructure are parameterized to exhibit higher costs to species movement. See Appendix B for additional details.

- **Transportation features, including Topologically Integrated Geographic Encoding (TIGER) roads and rail lines:** Roads and rails lines are provided from the US Census Master Address File (MAF/TIGER) database. These features are extremely common sources of edge habitat and often divide habitat cores.

- **Water features (National Hydrography Dataset, NHD):** Rivers, streams, lakes, and other water bodies are included. These data provide information on aquatic habitat and riverine systems, which often form natural corridors through landscapes.

- **Wetlands (National Wetlands Inventory, NWI):** The NWI is a publicly available resource that provides detailed information on the abundance, characteristics, and distribution of US wetlands. The NWI is provided by the US Fish and Wildlife Service. Note that this dataset may lack many smaller wetlands, and some are misclassified (e.g., standing water in the bottom of a rock quarry may show up as a wetland).

- **Terrestrial ecoregions:** Ecoregions are relatively large geographic areas that contain distinct types of natural species and communities. Developed originally by Olson and Dinerstein (2002) and Bailey (1995),[1,2] these data layers were modified by The Nature Conservancy (TNC) to be used in its biodiversity planning exercises as part of a process known as ecoregional assessments.

- **Soils (Soil Survey Geographic database, SSURGO database):** The SSURGO database contains information collected by the National Cooperative Soil Survey. This layer presents information on soils using map units, which are areas of soils with distinct properties and characteristics. Map units are organized as polygons, to which additional information can be joined from SSURGO's extensive database.

- **Land cover:**
  - The NLCD 2011 maps general land cover categories for the US and is the land cover dataset used to derive intact habitat cores.

- The National Gap Analysis Program (GAP) Land Cover maps detailed vegetation categories using the Ecological System classification system developed by NatureServe. Ecological systems are groups of vegetation communities that occur within similar physical environments and experience similar ecological processes.[3] Land cover categories are organized into a hierarchy with three levels, with Level One being the most general and Level Three the most detailed.

- The US National Oceanic and Atmospheric Administration (NOAA) has produced the Coastal Change Analysis Program (C-CAP) land cover dataset. This nationally standardized, raster-based inventory covers coastal intertidal areas, wetlands, and adjacent uplands for the coastal US data derived from the analysis of multiple dates of remotely sensed Landsat imagery. This dataset may not be included for states without any coastline.

- **Ecological system redundancy:** This is a raster layer that represents how many GAP Ecological Systems (Level 3) are found in each terrestrial ecoregion. A value of 1 indicates that the ecological system is found only within one ecoregion, meaning it has less redundancy.

- **Ecologically relevant landforms:** The ecologically relevant geophysical (ERGo) landforms dataset is a comprehensive classification of landforms based on hillslope position and dominant physical processes.

- **Ecological land units:** Ecological Land Units (ELUs) are areas of distinct bioclimate, landform, lithology, and land cover that form the basic components of terrestrial ecosystem structure. The ELU map for 2015 was produced by combining the values in four 250-meter cell-sized rasters using the ArcGIS Combine tool. These four components resulted in 3,639 different combinations, or ELUs.

- **Theobald Human Modification Index:** This raster layer estimates the degree of human modification of the landscape. Values range from 0 (least impacted) to 1 (most impacted). This is more fully described in *A general model to quantify ecological integrity for landscape assessments and US application*.[4]

- **Elevation:** This is a digital elevation model (DEM) from the National Elevation Dataset (NED). The NED is a seamless elevation dataset of the conterminous US, plus Alaska, Hawaii, and territorial islands such as Puerto Rico.

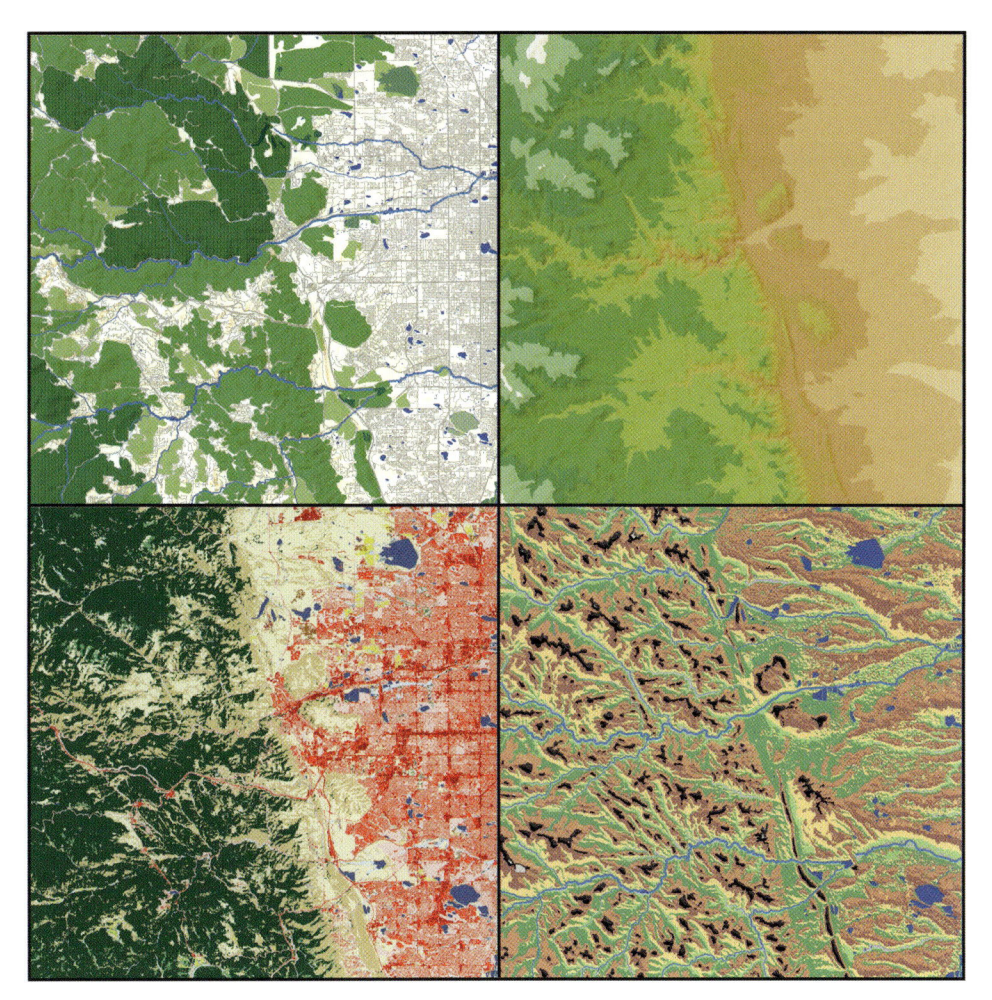

Important base data are included in state map packages, including data used to derive habitat cores and their associated metrics. Examples are shown clockwise from the top left: intact habitat cores, elevation, land cover, and landforms. Credit: Green Infrastructure Center Inc. Data from the National Land Cover Database, US Geological Survey; Esri; and the National Elevation Dataset, US Geological Survey.

The apps

There are several apps that help to access, filter, and prioritize habitat cores. These are particularly helpful in steps 2 (reviewing data) and 3 (making asset maps) of the GI planning process, as they enable the user to interactively explore the vastly diverse landscape characteristics represented in each core and provide content for the GI basemap.

Select the intact landscape cores

This app allows users to interactively select cores of interest and print or save them to their ArcGIS Online account, which can easily be shared with collaborators or used in additional mapping and analysis. The app allows users to zoom in to a city, county, or watershed and explore the habitat cores in view. The app will also select a subset of cores based on user-defined criteria. In so doing, users can refine and filter their cores to identify those that exhibit the landscape characteristics they aim to protect. For example, if biodiversity protection is a primary goal, a mapmaker can filter cores to identify those that have the highest total of endemic species.

Users of this app are leveraging a robust analytical process to identify priority cores; as such, the app is highly customizable and replicable. It can also generate a set of cores that embody the landscape characteristics a user deems important.

During the process, users can filter cores to select only those that meet certain criteria, for example, by size or stream density. The interactive nature of this process means that changes to the map are seen almost instantly, which is useful for a workshop setting where multiple users may be gathered to develop scenarios.

However, it may not be necessary to filter cores at this stage; instead, all the cores in an area of interest can be saved for later use in ArcGIS Online or the Score Core app.

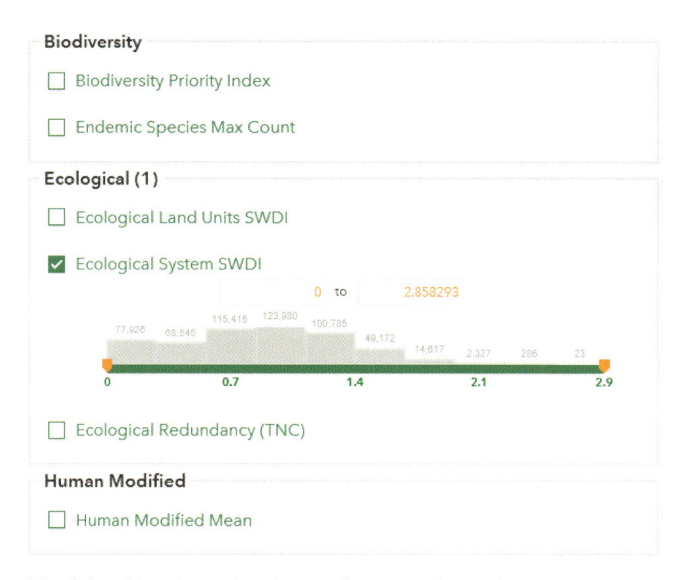

The Select Your Intact Landscape Cores app is used to query the national intact cores database to select cores of interest. Credit: Esri.

Prioritize the intact landscape cores

Next, the Score Intact Habitat Areas app can be leveraged to further refine the user's selection of cores. Here, the user can load saved cores from ArcGIS online and apply customizable weights to a selection of landscape characteristics from the first app. In so doing, the user can link original goals with the modeled landscape characteristics embodied in the cores. Again, if users have identified biodiversity protection as a primary goal, they could now place added weight/importance on the biodiversity priority index characteristic.

As a result, this app allows a user to quickly create ranking scenarios for the cores selected using the Select Your Intact Landscape Cores app. By creating ranking scenarios, the user can combine multiple variables to create an overall ranking for each core. If a user is not sure what ranking weights to use, recall that it is common to assemble a team of technical experts, including scientists who understand the ecology of the study area, to serve on a technical advisory committee. This app is an excellent tool to facilitate workshops with such a committee. Default weighting schemes are also available.

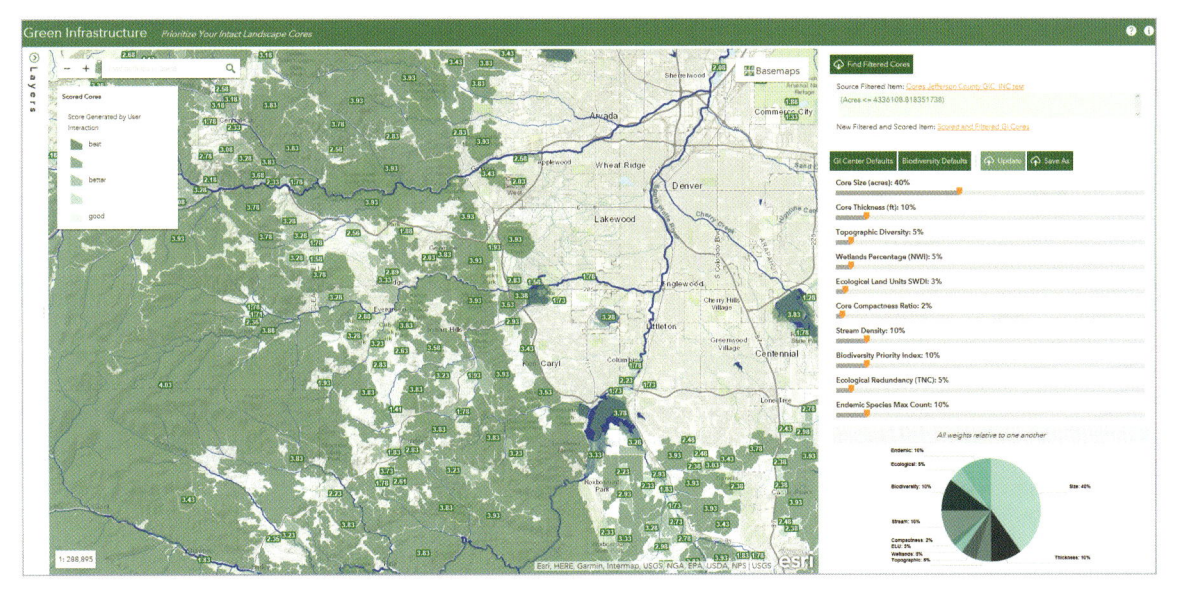

The interface of the Prioritize Your Intact Landscape Cores app allows users to rank and score intact landscape cores by weighting relevant landscape characteristics. Credit: Esri.

The model

The ArcGIS toolbox that generates intact habitat cores is available for ArcGIS Desktop and ArcGIS Pro. This toolbox allows advanced users to rerun the cores model. The most common reason for doing this is to incorporate local data into the model. For example, if a county has high-resolution land cover data available, you can use the toolbox to rerun the Cores model using the higher-resolution data as a substitute for NLCD 30-meter data. This has the potential to produce finer-grained cores data.

Downloading the tools also allows advanced users to modify model parameters and assumptions.

Which method to use?

Downloading a map package is the fastest method to obtain data using ArcGIS Desktop. You can download data for entire states, including habitat cores data and many supporting datasets. All datasets are symbolized in the map package, which can save time when building those maps.

Example use cases:

- A GIS analyst who is tasked with assisting with a GI plan
- A land trust with GIS capacity that is seeking to inventory the natural assets in its region
- A student or researcher who is collecting data for a city, county, or state

Use the web apps

The Select Your Intact Landscape Cores app and Prioritize Your Intact Landscape Cores app make it easy to explore the cores data, even for those without ArcGIS desktop software. Non-GIS users can quickly zoom to their area of interest and view the location of cores, as well as explore the wealth of attributes calculated for each core. The apps also make on-the-fly filtering and ranking possible, which is useful for exploratory analysis or facilitating a workshop with an advisory committee.

Example use cases:

- A non-GIS user who wants to quickly view the assets in his or her area
- A GIS analyst who is tasked with facilitating a meeting with a GI advisory committee, including landscape ecologists, wildlife ecologists, and other natural resource experts; the apps are used to create ranking scenarios on the fly
- A GIS user who wants to seamlessly consume the cores data in his or her organizational ArcGIS Online account and/or Geoplanner applications

Download the toolbox

The toolbox allows advanced users to run the model that generates habitat core data. This may be useful for those who wish to change model parameters or substitute local data in the model.

Example use cases:

- An advanced GIS analyst who wishes to identify habitat cores in his or her locality, but instead of using national data (NLCD, TIGER roads data), he or she uses the 1-meter resolution land cover that the locality has available
- An advanced GIS user who wants to build a custom habitat model and uses the toolbox as a starting point

Limitations of the habitat cores data

As discussed earlier, the habitat cores data are most appropriate for use in strategic, landscape-scale planning. The most significant limitations of the data originate from the land cover data used to identify habitat cores. The cores are identified from land cover data from the NLCD 2011, which is produced at a resolution of 30 meters and was last updated for the contiguous US in 2011 (a new update is expected around the time of this book's release). Smaller physical features, such as a small patch of trees, may not be represented in the source land cover data. Thus, the data will be of limited use at a site scale or in urban areas. However, you can use the data to place smaller assets in context or identify opportunities to connect to a larger network.

Changes to land cover since 2011 will not be reflected in the cores data. Users should keep in mind that the Esri model may be rerun in the future with new land cover data from the next update of the NLCD. Therefore, check the date of the latest NLCD.

Adding local data: Creating customized maps to reveal relevant priorities and guide actions

Users should recall that a map of intact habitats is not a GI plan. To create such a plan, you should incorporate community goals as a driving force, as described in chapter 3. Likewise, you should include community data wherever possible to make the GI plan more accurate and locally relevant. Often, the best starting point for gathering such data is the GIS department within a city or county government or a regional planning agency.

Note that unique datasets may also be housed in other departments or government agencies. Consider the following scenarios:

- **Scenario 1:** A city parks and recreation department has GIS data related to a recently completed Trails Master Plan that show proposed trails and infrastructure upgrades or locations for future open space. The contractor who completed this plan delivered the project's geodatabase to parks and recreation directly, and therefore the GIS department does not have the data.

- **Scenario 2:** A county's water and sewer authority maintains data that show the areas served by the water and sewer system and planned expansions to the water and sewer network. Utility service areas are extremely useful when assessing the risk that development poses to natural lands, because the ability to connect to a water and sewer system is usually a prerequisite for moderate-to-intense development.

- **Scenario 3:** The GIS department in a locality has basic parcel information, but contacting the tax assessor's office is necessary to get records that indicate whether a parcel is included in the locality's Use Value Assessment Program. The tax assessor provides the records in tabular format that can be joined to the parcel GIS layer. Many localities have programs that provide tax relief to landowners who use their land for certain agricultural, forestal, or open space purposes. This information can be used to better understand where agriculture and forestry are taking place, although tax status does not always reflect the actual use of the land.

Thus, you often must search beyond the GIS department to collect the data required to plan for GI. Geodesign, mentioned in chapter 1, is a planning process for bringing in multiple disciplines and public participation to ensure thoughtful plans that meet local needs.

Recall the three primary reasons to GIS data for GI planning, as described earlier. Datasets available from local government GIS departments are especially useful for the second and third reasons (assessing risks and providing reference). For example, local governments maintain many layers related to land development, such as zoning and future land use, which can pose risks to habitat. Local datasets can also help identify GI assets, but often these assets are not commonly maintained at the local level. For example, habitat cores data and prime agricultural soils are not commonly found in a local government GIS but are key natural assets.

Filling data gaps

Most localities have not performed their own soil survey or delineated their own wetlands. You should use local surveys if they are available. When local data are not available, explore data available from state, federal, or regional agencies. State agencies that deal with natural resources typically maintain and provide data. These agencies may include departments of conservation, recreation, agriculture, and forestry. When considering whether to use a dataset in a map, use the following three general rules:

1. The data must exist (or be available in the near term). Revealing deficiencies in data is a common occurrence when planning for GI, as the scope of data collection is quite large. This by-product is useful in guiding future data collection efforts in the planning process. However, it should not prevent GI planning. No locality in the country has a complete survey of all plants and animals within its boundaries. But all conservation actions cannot wait until perfect information is available. Rather, use the best current data available to make the most informed conservation decisions possible, because development will most certainly continue in the absence of perfect information. Some types of data may not be readily available but can be easily collected. For example, datasets showing places of historical significance often underrepresent the histories of African Americans, Native Americans, and other minority groups. You can help address

this issue by bringing a map to local museums or universities and allowing subject experts to add their knowledge to the map. You can hold community meetings to learn of locally significant sites.

2. The data must be represented spatially. In other words, the data must be able to be placed on a map. Data already in a GIS format obviously meet this criterion, but remember that you can also map tabular data as long as location information is attached to each record (such as latitude/longitude or an address column in a spreadsheet). You can collect a list of historic buildings in spreadsheet form, for example, but you can turn this list into spatial data quickly by geocoding the addresses. Conversely, a simple list of the rare, threatened, and endangered species data found in a county cannot be mapped, as these data usually lack location information and therefore have limited usefulness.

Instead, try collecting data from a state's natural heritage division, which details where these species have actually been observed. This collection method allows you to include the geographic distribution of these species in a data layer, where it can inform the prioritization of one area over another. If you are concerned about identifying locations of rare and endangered species or vulnerable monuments and cemeteries, you can buffer each site to disguise the actual locations. Buffering locations can help discourage poachers or treasure hunters from hunting rare species and looting artifacts.

3. The data must be consistently available over the entire study area. In some cases, a detailed survey may have been completed for a site within the study area. However, some caution is advisable when adding such localized survey data to a map. The information may be appropriate to inform later stages of prioritization, but it may be inappropriate to use such data in a weighted overlay. For instance, when habitat cores are being compared directly with one another, it is best to use consistent data across all ranking factors.

The Living Atlas of the World

Esri maintains the ArcGIS® Living Atlas of the World as a collection of authoritative geographic information from organizations around the world. The Living Atlas contains live maps and layers that you can access in a map or model, as well as downloadable data and models. The live layers are a simple way to fill gaps in data and map natural and cultural assets. On the Esri Green Infrastructure website, datasets and layers from the Living Atlas are categorized to facilitate GI planning. Categories include Ecological Assets, Cultural Assets, Scenic Assets, Hazard Zones, Conservation, and Boundaries.

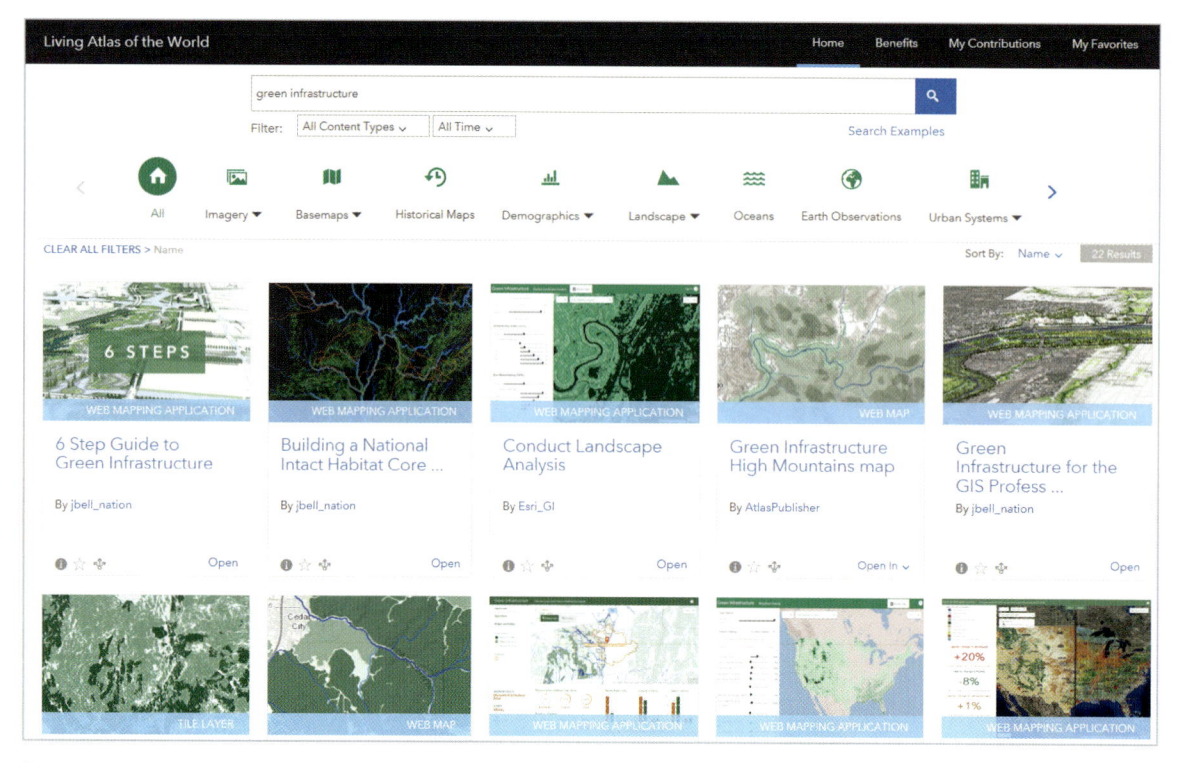

Data resources in the Living Atlas can be used to identify GI assets or risks to those assets. Credit: Esri.

Data surrogates

You may encounter situations in which no data are available from local, state, or other sources. In these cases, you may find another dataset as a surrogate to achieve a similar result. These data are known as *proxy data*. For example, data on historic buildings may be extremely limited or even nonexistent for a study area. One option would be to pinpoint older structures using parcel data if there is a "Date Constructed" field in the attribute table.

Consider that many datasets used in modeling are already proxies. For example, habitat models may use proxy data to estimate where animals are *thought* to live, not where they have been observed. Similarly, a development pressure model may use "distance to major roads" as a proxy for development pressure, because actual development pressure is difficult, or impossible, to measure definitively.

Data scales

Mapping GI assets requires the combination of datasets that were created for a variety of purposes, and thus were created with a certain map scale in mind. Using this information is not generally a problem when creating GI maps if the limitations of the input datasets are recognized and noted in maps, analysis products, or reports. When collecting data, consider the following two questions:

1. Can the dataset inform the prioritization of one area over another? For example, you can overlay a state department of natural heritage dataset that depicts the locations of rare, threatened, and endangered species on to the Esri Habitat Cores model to determine which cores may contain those species. You can use this metric as an additional factor when prioritizing one core over another in land-use and conservation decisions.

2. Can the dataset provide context for the study area? For example, a county may obtain a dataset that depicts a migration zone for monarch butterflies. However, the entire county is shown to be within this migratory zone. In this case, the data do not help prioritize habitat cores within the county but *do* provide context for the county. Such context can inform more general GI strategies, such as encouraging residents to create pollinator gardens full of nectar flowers for butterflies and food sources for caterpillars. In this example of a nonspatial strategy, no specific locations are mapped.

Moving on to mapping

The data collection phase of any GIS project can be daunting. GI planning requires the use of a variety of datasets from multiple sources. This requirement can create new challenges, because agencies traditionally maintain and keep the data within their respective agency. In contrast, GI plans cut across many areas—conservation, agriculture, forestry, water supply and quality, parks and recreation, and so on. Thus, a main strength of the GI planning process is combining disparate datasets (and people) to create new insights.

Although data collection can be a challenging (and sometimes tedious) phase, the level of effort required can be reduced significantly if it is as targeted as possible. As discussed in chapter 3, establishing goals for the project is the first step in the planning process. To a large extent, these goals will drive subsequent data collection efforts. For example, if a community is a popular destination for birders, one goal might be to disseminate information and enhance infrastructure related to birding. This would require collecting spatial information on popular birding locations, potential birding locations, and the location of bird blinds.

Although a community's goals drive much of the data collection, the process remains iterative; the need for new data will likely occur throughout the course of a GI planning project. One of the best ways to identify needs is to use an advisory team that includes multiple representatives from departments that will be involved in the implementation of a GI plan. For example, Hot Springs, Arkansas, created a "Green Team" made up of representatives from the departments of planning, public works (stormwater), and parks and recreation and several elected officials. This team served as a working group for the city's GI plan. The group identified the city's goals for the project and reviewed maps and data throughout the process. The team also provided initial input on data that were and weren't available across their departments. This work, in turn, determined whether planners needed help from outside groups for datasets that the city did not maintain, such as the National Park Service, which operated sites within and around the city.

How can the new cores data be used to inform land-use decisions? The answer is that you can access these data in many ways to see whether current policies or proposed changes will impact intact habitat cores, a foundational piece of every region's GI network. The following case study is from the Piedmont region of central Virginia, a mountainous area of rolling hills bordered by the Blue Ridge Mountains to the west.

Case study: Using habitat cores to assess zoning changes in Albemarle County, Virginia

Albemarle County, Virginia, is divided into a rural area and a development area, as specified in the county's comprehensive plan. The development area is for residential and business growth and priority investments in new infrastructure (e.g., water, sewer, sidewalks). The rural area complements the development area by emphasizing the county's rural character, including thriving farms, wineries, forests, and Shenandoah National Park on the western border. The rural landscape makes up most of the county's land. The rural economy is of tremendous importance to the county, so it has an active land conservation program and sets aside funds through the Acquisition of Conservation Easements (ACE) program to acquire development rights that protect areas important for environmental benefits, agriculture, or tourism. As of 2017, 121,000 acres of the county's 725 square miles of land were protected as parkland and by conservation easements.

As the county's unemployment rate is quite low at 3.5 percent, and a major interstate highway runs through the middle of the county and the University of Virginia in the area, the landscape faces demands for development as the population and economy grow.

In 2015, a large brewery from the West Coast identified the county as one of three possible locations for a new brewery and tasting room. The county lacked sufficient industrially zoned land with topography and utilities to support a new factory. The county, therefore, looked to its highway interchanges to find a site that could be rezoned for the brewery. Some community members asked, "How would the new business impact resources?"

The next figure in this chapter shows that rezoning and clearing the land would affect prime core habitat. As the project unfolded, the county did not yet have a map of habitat cores, such as the Esri National Intact Cores model. The county proposed to rezone 137 acres from rural lands to development lands to accommodate the land requested by the brewery and additional lands in case any other factories or commercial applicants applied to use that land in the future. A map did not exist to show the locations of core locations in the area proposed for growth.

Community concerns arose about whether county residents would fill the new jobs created, because it would be necessary to commute from surrounding areas. In addition, some residents opposed the loss of a large habitat area containing mature forests close to the developed area. The area was also adjacent to a new park recently donated by The Nature Conservancy, which would also have been affected by the new development.

After many meetings with the community and with local government officials, the local planning commission rejected the project for reasons that included requiring the removal of a large forest with many mature trees. The county ultimately reduced the size of the rezoning from 137 acres to 25, but the brewery located elsewhere, deciding the area lacked infrastructure and transportation and was not "development ready."

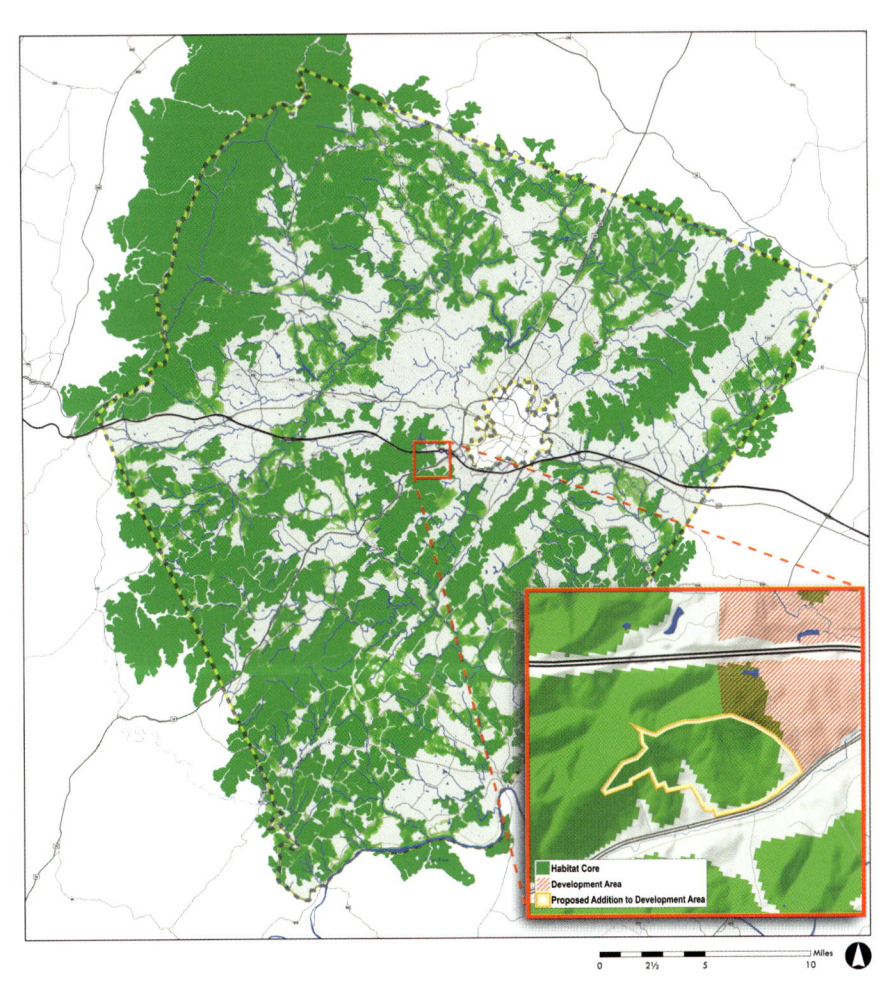

Land that was proposed to be rezoned for development. Credit: Green Infrastructure Center Inc.

Chapter 4 | Getting the right data

This case study shows why access to data is important. Knowing from the start that about 120 acres of core habitat would have been affected could have armed citizens and the planning commissioners with the information they needed to resist the project. Since then, the Green Infrastructure Center has helped the county create a detailed map showing where all its key cores are located. With this information, the county can better plan future development.

The county still may decide to develop some of the cores. But now the county knows the locations of preferred development areas (because they have less valuable habitat and meet other goals) as well as areas that should be preserved (because they have important habitat cores and high infrastructure costs or protect local waterways). In 2018, the county adopted the new habitat cores maps as part of its Biodiversity Action Plan, which was also adopted as part of the county's Comprehensive Plan. This case study shows why maps are so important: they can provide objective information about what resources are present and where they are located—*before development plans are made or land is rezoned*. Areas with sensitive resources can also be protected through additional overlays, such as a conservation protection overlay that may include additional provisions that protect rare, sensitive, or scenic resources. Any county in the US can access this data through Esri's national green infrastructure model.

This chapter covered how to obtain the right data, add local data, and customize maps for the community, along with tips for using apps developed by Esri for viewing the GI data. Chapter 5 will explain the process for creating customized maps and working at the right scales for decision making. The chapter will also cover options for making maps that identify the most valuable GI assets in a community to meet local and regional goals. Getting data is a key step, but knowing how to represent the most useful information for decision making is central to an effective GI plan.

Notes

1. Olson, David M., and Eric Dinerstein, "The Global 200: Priority ecoregions for global conservation." *Annals of the Missouri Botanical Garden* 89, no. 2 (2002): 199–224.
2. Bailey, Robert G., comp., *Description of the Ecoregions of the United States*, 2nd ed. Misc. Publ. No. 1391 (Washington, DC: US Department of Agriculture, 1995).
3. Aycrigg, Jocelyn L., Anne Davidson, Leona K. Svancara, Kevin J. Gergely, Alexa McKerrow, and J. Michael Scott, "Representation of ecological systems within the protected areas network of the continental United States." *PLOS ONE* 8, no. 1 (2013): e54689.
4. Theobald, David M., Sarah E. Reed, Kenyon Fields, and Michael Soule, "Connecting natural landscapes using a landscape permeability model to prioritize conservation activities in the United States." *Conservation Letters* 5, no. 2 (2012): 123–133.

Chapter 5

Making asset maps

Chapter 4 covered how to obtain the right data for decision making. Having the right data is just one step in the GI planning method's six-step process. This chapter explains the third step for mapping natural and cultural assets. Note that taking the first step (setting goals) and second step (getting the right data) are critical before you take the third step (making asset maps). GIS users know it is fairly easy (and becoming easier) to produce maps, but the object of this step it not to produce as many maps as possible. Rather, the purpose is to produce a series of targeted, useful maps that inform the goals in step 1.

A GI landscape map should reveal potential opportunities to make the goals a reality and expose potential barriers and constraints. For example, if a goal is to preserve biodiversity, then *where* can you find the richest intact habitat cores? If a goal is to protect water quality and supply, then *where* can you best focus on aquifer recharge and riparian buffers? You must know the locations of the best natural and cultural assets to effectively protect and enhance them.

The quality of water in this old quarry in Ulster County, New York, is protected by the forest cores that surround the quarry. Credit: Green Infrastructure Center Inc.

Making asset maps also discloses where *multiple* benefits are being created (or could be created). For instance, if a forest surrounding a reservoir preserves clean drinking water while also preserving intact habitat, multiple benefits are safeguarded. Assessing each benefit in isolation may not justify protecting the forest, but showing multiple values may provide a compelling argument for doing so. Identifying locations where multiple benefits are met can provide a powerful case for conservation or restoration.

A multiple, interconnected benefits approach

Taking a *multiple, interconnected benefits* approach can show how GI is a complex network of mutually interactive assets that can be a mix of both cultural and environmental features. For example, a rural landscape might support a network of plants and animal species, as well as provide a historic viewshed and enhance cultural assets, such as an 18th-century mill or stretch of historic canal, by buffering them from impacts or providing a natural setting in keeping with past views.

The landscape that surrounds a site provides the context for that site and might even be part of the story that makes the site unique. The views looking away from a site are normally crucial to preserving its cultural context and often provide habitat for a multitude of species. For example, Monticello, the home of President Thomas Jefferson in Charlottesville, Virginia, provides not only a key cultural site but also the forested views that radiate for miles across neighboring hills, part of its appeal and historic context. Several grand houses in the Southeast, notably Magnolia Plantation and Gardens near Charleston, South Carolina, preserve bird sanctuaries and protect threatened marshland, old rice paddies, and stands of old bald cypress. Similarly, outdoor recreation—whether mountain biking, birding, or rock climbing—is especially appealing because of, for example, protected bald eagle nesting sites or the presence of bighorn sheep.

The Buffalo River flowing through Ponca, Arizona, provides the scenic vista that supports a multitude of recreation uses. Credit: Green Infrastructure Center Inc.

The process of mapping natural assets and then adding overlays of cultural assets, environmental resources, and outdoor sporting activities provides the opportunity to look at those specific assets in relation to each other and to a common framework.

Often, certain assets will appear on the same map for the first time using this process. For example, counties may map their best agricultural soils as part of an agriculture-specific project and show soils regardless of overlaying land uses. Thus, some soil types might be under a town, a suburban subdivision, or state forest. Mapping prime agricultural soils as part of a GI planning project shows incongruities in relation to key habitats and water resources, as well as recreational and development areas, among other things. The mapping highlights where agriculture takes place, where it could take place, where it could provide multiple benefits, and possibly where such lands are at risk. It also firmly places agriculture within the context of adjacent land uses, which can be important to visualize and understand for future planning purposes. A forest buffering a farm operation may provide a key landscape screen for current housing developments or filter farmland runoff. However, agricultural soils underlying a forest should not appear on an agricultural map overlay as available agriculture land, because it is unlikely that the forest will be cleared for agriculture.

Who makes the maps?

Asset maps are created by someone proficient in both GIS and cartographic best practices. Making such maps typically requires manipulating and presenting data clearly. A GIS analyst or project team member should fill this critical role. For this chapter, we refer to this person as the GIS lead. A GIS lead should be a core member of the project team. If, for example, the GIS lead is an analyst within the GIS department, she should attend the team's goal-setting meetings. Otherwise, she would have no frame of reference for the maps that must be made, and the mapping process would become yet another dictated task. Staying involved in *all* project team meetings will help the lead better understand how the goals were created and what the maps are supposed to communicate. Full involvement in the process will help the GIS lead communicate effectively with maps.

An iterative process

As explained in an earlier chapter, project leaders should remember that asset mapping is an iterative process. The GIS lead produces draft asset maps, brings them to the project team and other reviewers to assess their accuracy, updates the maps, and brings them back to the team for review as needed.

If special expertise is needed, the GIS lead can facilitate the map review by bringing in subject matter experts and forming a committee of experts. For example, a GIS lead could gather input from local historians or from the local historical society This expertise is helpful because GIS data for historic sites are often incomplete and limited to government datasets, such as the National Register of Historic Places. A GIS lead might also seek expertise from sporting clubs, such as birders' associations, wildlife protection agencies, and hunting and fishing clubs to gather information on local plants and animal species and their habitats.

Project teams often rely on advisory groups of local experts to suggest additional cultural sites for mapping. For example, Native American and African American sites often were omitted from maps in the past. Other sites not previously regarded as significant—for example, a row of mill worker houses once regarded as an eyesore—now may be valuable to preserve as cultural history. In one case, a Civil War memorial covered an auction block for slaves. In this case, the auction block was mapped for the first time.

Similarly, a scientific advisory group of botanists, ecologists, foresters, and other experts might advise on locally known rare plants or animals. A particularly complex subject might require input from an advisory group of experts from multiple disciplines.

Working with technical experts: Why have a formal advisory group?

One of the strengths of GI planning is that its data create realistic strategies for managing a community's best natural assets. However, GI projects incorporate a lot of information on various topics. For this reason, it often helps to create an advisory group to work on a project's technical and scientific components. This group might, for example, help the GIS lead interpret soils data and relate them to agricultural value or land cover types. The group might determine a minimum core size for a specific habitat, such as Carolina Bays, that provides adequate habitat for local species and allows planners to select bays or islands that are most crucial to species diversity.

Mapping Carolina Bays: Method case study

Carolina Bays are elliptical depressions in the ground that occur in many places in the eastern US and are noted for being particularly rich in biodiversity. Varied in size, Carolina Bays are usually wetlands. They can span a spectrum from mostly open water to mostly scrub vegetation and are found in both North and South Carolina. During a GI project in Darlington County, South Carolina, Carolina Bays were identified as an important natural asset.

As Carolina Bays had not been specifically mapped, they could be excluded from an asset mapping process, even relegated to a footnote in the final report. However, looking for opportunities to tease out additional insights from currently available data is an important part of mapping natural assets.

Fortunately, Carolina Bays tend to have several distinguishing characteristics that can help identify them in a landscape. They are easily visible on aerial photos, even if they have been drained, cleared, and cultivated (as the area is elliptical in shape and will often appear darker because of the soil type). Three characteristics were identified that could be mapped:

- **Soil type:** In consultation with a soil scientist, it was determined that Carolina Bays, in Darlington County, are most often of the Rains or Coxville soils series. These soil types can be queried from publicly available SSURGO survey data.

- **Compactness:** A compactness ratio was calculated in GIS using SSURGO soil survey polygons to compare the area of each polygon with the area of a hypothetical circle whose circumference is the same linear distance as the polygon's perimeter. Essentially, this gauges how compact a polygon is by comparing it with a circle, which is the perfectly compact shape. Since an ellipse is a relatively compact shape, this is another metric that can be used to identify Carolina Bays.

- **Orientation:** In addition to being elliptical, Carolina Bays tend to orient northwest to southeast (along their long axis). The ArcGIS Minimum Bounding Geometry tool assessed this information to create the minimum bounding rectangle around each potential Carolina Bay, and the option for adding geometry characteristics to the output feature class was enabled. The tool then added the orientation of each bounding rectangle to the output attribute table. In this way, a relatively compact polygon could be ruled out as a Carolina Bay if its orientation was very different from what would be expected.

The GI project identified polygons that exhibited the key characteristics as likely Carolina Bays, and a manual review of the results against aerial photography was performed as a final check. However, many Carolina Bays have been cleared of their natural vegetation, mostly for agricultural purposes. Thus, a final overlay was done using land cover data to identify which polygons identified using these metrics were still mostly intact. Polygons containing greater than 50 percent forest or wetland land cover were separated to be included in GI asset maps.

This methodology has limitations. For example, small Carolina Bays may not be captured as a discrete polygon in the SSURGO soils survey data, in which case it would not be identified by this method. However, this method produced a new dataset identifying the location and extent of likely Carolina Bays relatively quickly using publicly available data. These data were later used to inform the overall ranking and prioritization of habitat cores (does a core include a likely Carolina Bay?).

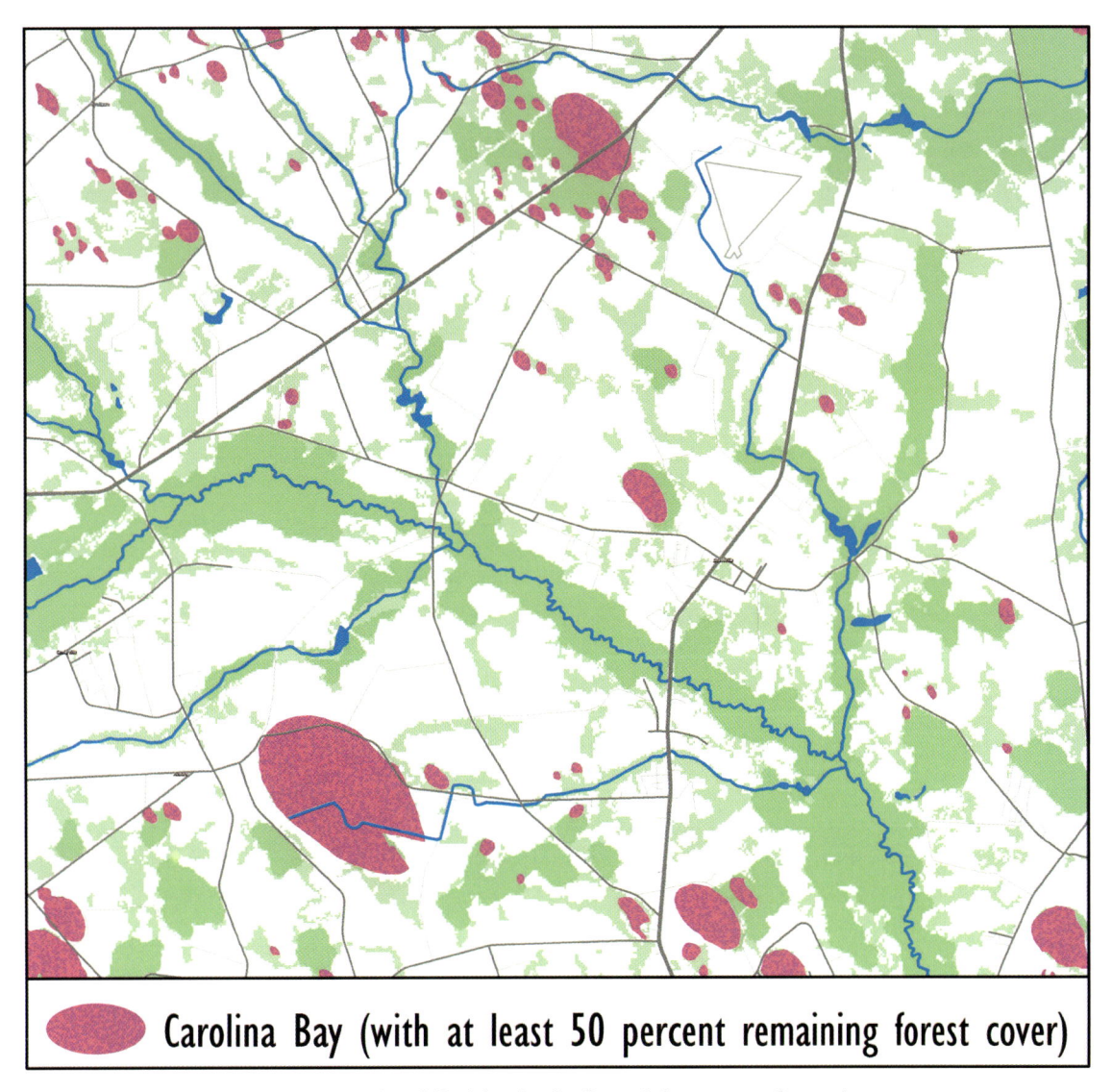

Carolina Bays can be revealed using a few GIS tricks. Credit: Green Infrastructure Center Inc.

Following are examples of ways in which a technical advisory committee can be involved at key steps in the GI planning process:

- **Step 2**: Review data: A science/technical advisory committee can help identify useful datasets or studies that can be mapped or used to inform strategies.

- **Step 3**: Make asset maps: A science/technical advisory committee can determine mapping rules that expand upon the goals identified in step 1. For example, if one goal is to "provide connected habitats that support biodiversity and healthy landscapes," then the advisory group can determine mapping parameters to identify key areas for maintaining habitat connectivity.

- **Step 4:** Assess risks: A science/technical advisory group can estimate the impact of certain actions and inform the development of alternative scenarios. Its members can also ensure the diversity of knowledge and perspectives for decision making. For example, they may point out how fragmenting different habitats across a landscape would negatively affect the overall health of an ecosystem or where siting new development on productive agricultural and forestal soils would affect the rural economy.

You can include a variety of stakeholders and experts in a review committee. To identify potential members, first consider the types of goals and data used in the planning and analysis stages. Do the goals include preserving unique and productive soils? If so, you can add a soil scientist or expert from an agricultural extension service to inform this goal. Do goals include protecting the community's most sensitive habitats? In this case, a wildlife biologist or landscape ecologist from the state's natural heritage program or a local university can help the GIS lead develop a ranking scheme for habitat cores based on locally important criteria.

Often, such expert members of an advisory committee are affiliated with local, state, or federal agencies involved in managing natural assets. Consider which agencies have a regional office that has experts knowledgeable about local and regional conditions.

As noted earlier, experts can lose sight of funding and implementation realities at times. They might recommend extensive field studies or research that lacks funding, so you must balance their ideas against practical constraints. Advisors will likely need an orientation before offering advice, especially if they are unfamiliar with cores and corridors or habitat modeling.

Creating the basemap

As described in chapters 2 and 3, creating a basemap is the first major mapping task. A basemap identifies intact habitat cores, connections across the landscape, and major landscape elements such as topography and water features. The basemap provides context and other GI assets and a common framework for overlaying and comparing them.

Mapping habitat cores

You can use an existing habitat model, such as a statewide model, to start mapping habitat cores. Chapter 1 noted several examples, or you can use the Esri intact habitat cores layer, because it identifies large areas of intact natural land cover and can act as the backbone of a GI network. However, mapping the location of habitat cores is just the start. To develop informed conservation priorities, you must explore the metrics of each core that makes it unique.

In the Esri cores dataset, for example, a variety of metrics have been calculated that describe each core's metrics, including the following:

- **Geometry:** Size, shape, interior depth, and perimeter-to-area ratio
- **Physical characteristics:** Topographic diversity, landform diversity, and dominant land cover
- **Ecological characteristics:** Dominant ecoregion and ecosystem rarity

Mapping these metrics, or a combination of them, informs the prioritization process. Thus, even though habitat cores are themselves a subset of habitat in a given area, creating a prioritization scheme based on core metrics is useful to further refine priorities on the basis of local goals.

To show how this works in practice, the authors obtained data for Jefferson County, Colorado, and used the Esri model as a basemap for the cores. The county is home to about 395,000 acres of natural land, or land cover not classified as developed or in agricultural use. Within this acreage, 330,000 acres (83 percent) are in large enough and intact enough tracts to be considered habitat cores—some 110 habitat cores in total.

The questions become how to preserve and prioritize these cores, given other pressures on the land. The first step is to rank cores depending on their characteristics and how well those characteristics conform to the community's stated goals. As it is unlikely that everything can be preserved, further analysis is necessary to uncover these priorities to ensure that strategic lands with the highest value, in terms of desired metrics, are protected when difficult choices must be made.

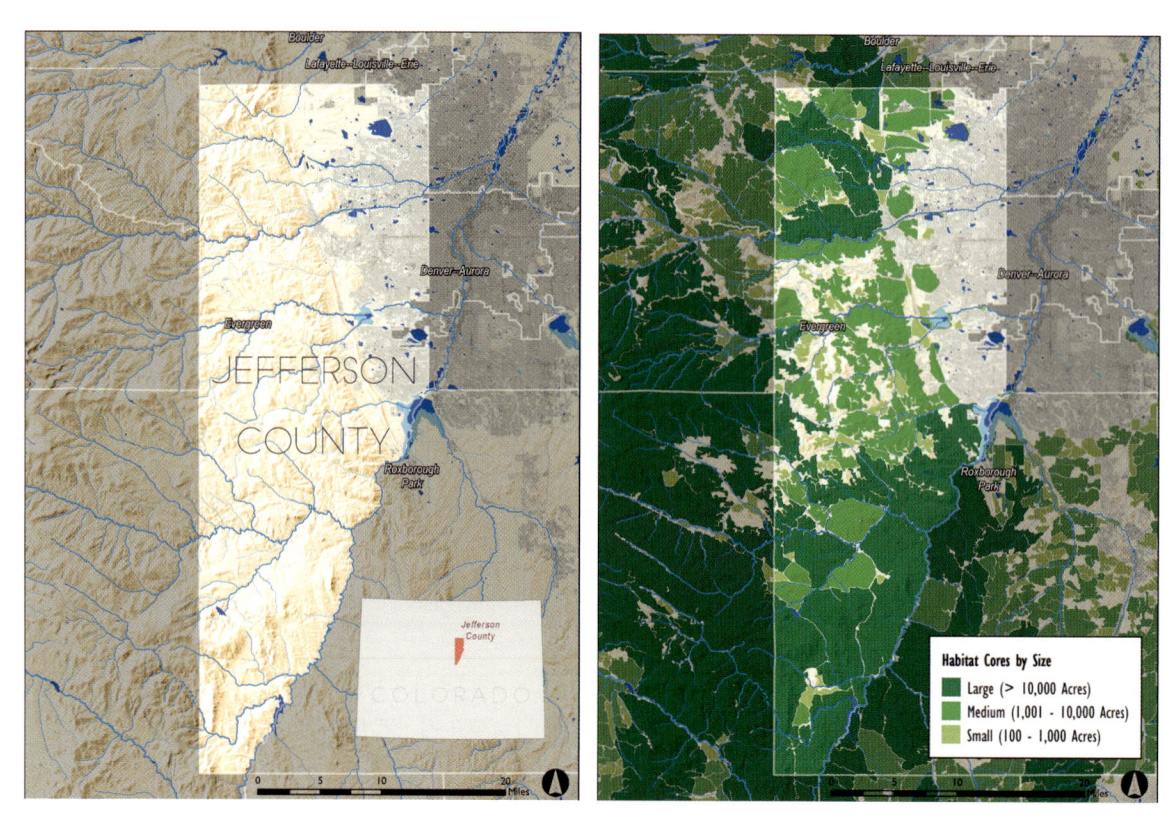

At 77,000 square miles, Jefferson County is a large county considered the "gateway to the Rockies." The map of Jefferson County, Colorado, shows habitat cores categorized by size, which is often the single best predictor of ecological integrity. Categorizing cores by size (using a variety of different symbols or colors) is a simple and effective way to display the data. Credit: Esri Green Infrastructure Model.

Ranking habitat cores

The intact habitat cores layer provided by Esri contains many metrics that describe each habitat core. These metrics can be mapped in any of three ways:

- **As individual metrics:** For example, it can be a map of cores, symbolized by size from light green to dark green.

- **As multiple metrics:** These metrics can meet a wider complex of criteria, for example, a map showing cores that contain a rare species and are in the top quartile of landform diversity.

- **By creating a single score:** Multiple metrics can be collated into a single score, often weighting individual metrics based on their perceived importance. This score can serve to summarize multiple metrics.

Several example summary scores have been calculated for the Esri cores dataset and are included in the attribute table. This example used 10 metrics to create the summary scores in the Esri Habitat Cores dataset:

1. Core size in acres.

2. Core thickness, which measures the depth to the interior of the core. It identifies the radius of the largest circle that can be inscribed within the boundaries of the habitat core's polygon.

3. Topographic diversity, according to the standard deviation in elevation.

4. Presence of wetlands, by percentage of core area.

5. Ecological land unit diversity, using the Shannon-Weaver index.

6. Compactness ratio compares core area relative to the area of a circle with the same circumference as the total perimeter of the core polygon.

7. Stream density, in linear feet of streams per acre.

8. The Biodiversity Priority Index is measured by overlaying the intact core areas on the Priority Index layer (10 kilometers) resolution surface described in "US Protected Lands Mismatch Biodiversity Priorities" by Jenkins and colleagues.[1] The Priority Index score is a summary for 1,200 endemic species of the proportion of their unprotected species range divided by the total area of that range:

$$\frac{\text{Unprotected range}}{\text{Total protected range}}$$

9. Values are summed across all endemic species within a taxonomic group and across all taxonomic groups. Cores falling within a priority index category are assigned that priority index value. Note that the nominal resolution of the Priority Index data is 10 kilometers. Cores may or may not have endemic species or collections of endemic species within them.

10. Ecological system redundancy measures the number of TNC Ecoregions Systems in which a GAP Level 3 Ecological Systems occurs. The higher the number, the more ecoregions in which an ecological system appears and the greater its redundancy. Cores are scored according to the lowest redundancy value appearing within them. Low and very low redundancy values represent cores containing unique ecological systems.

11. Endemic species max records the number of endemic species whose entire range falls within the contiguous US and whose range intersects the habitat cores being surveyed. Cores that contain the ranges of a greater number of endemic species will have higher values.

To create a composite ranking that incorporates all 10 metrics, cores are assigned a value from 1 to 5 for each metric. These values are assigned according to the quintile within which each core falls, with 5 reserved for the top quintile and 1 reserved for the lowest quintile. For example, if a core is large enough that, when all cores are grouped into five equally sized groups after being sorted from largest to smallest, it falls into the top category, it would receive a score of 5 for that metric.

> *The method of ranking GI cores is an example of a* **relative ranking scheme** *in which the cores are ranked against each other, not based on hard thresholds. Thus, the final score of a given core depends on the characteristics of the cores against which it is being ranked.*

After a score has been calculated for each metric, it is multiplied by a weighting factor, and then all the scores are averaged. The weighted average of the 10 metrics makes up the final composite score.

Two versions of this composite score are included in the habitat cores dataset:

1. **National ranking:** Cores are ranked relative to all other cores in the nation.
2. **Ecoregion ranking:** Cores are ranked relative to all other cores in the same ecoregion. This ranking allows every core to be compared with all others with the same climate and geology.

In addition, two modifiers are calculated for both the national ranking and the ecoregion ranking:

1. **The National Hydrography Dataset (NHD) Plus modifier:** Before being ranked (either nationally or by ecoregion), cores that do not include a stream with a least 1 cubic foot per second of estimated annual flow are excluded. This score is calculated for only a subset of cores that meet this NHD Plus flow estimate, based on the NHD). This modifier may be of use in more arid environments where the presence of water is of special concern.
2. **High biological component weighting:** This modified score changes the default weightings for ranking metrics by increasing the relative weights of the biodiversity priority index, ecological system redundancy, and endemic species max components of the composite score. Specific weights are detailed in Appendix B.

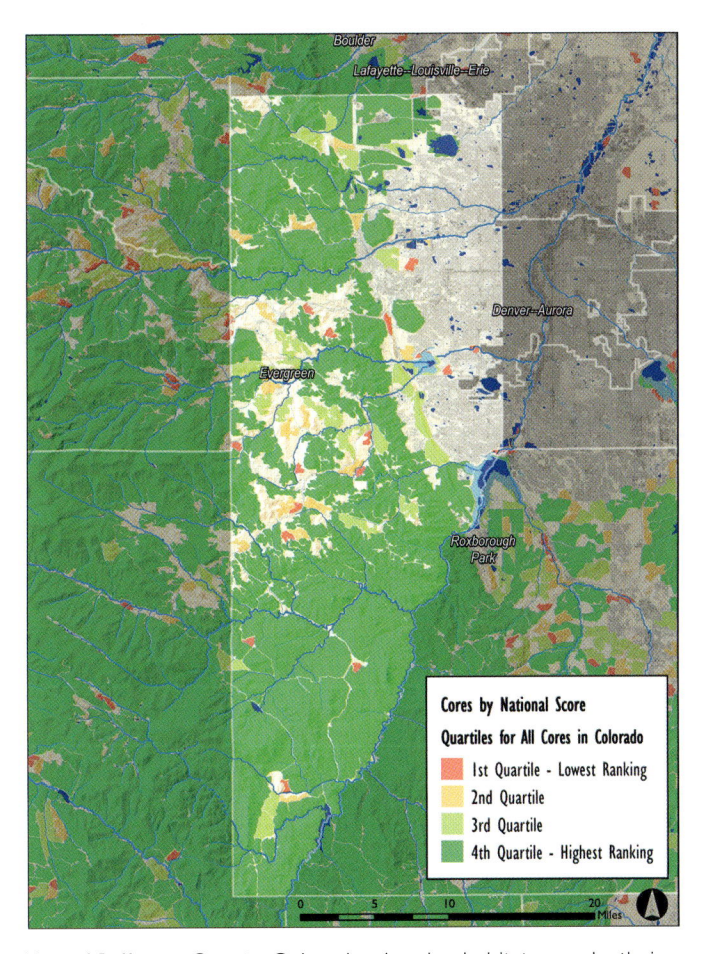

Map of Jefferson County, Colorado, showing habitat cores by their national ranking. In this scheme, cores are ranked against all other cores in the contiguous US. This composite ranking is displayed using quartiles based on habitat cores in Colorado. Credit: Esri Green Infrastructure Model.

Turning data for habitat cores into a visual format

Even though habitat cores data can include a wealth of information associated with each core, the method used by the cartographer to format that data for visual display will influence how the maps are interpreted. Typically, static maps are the primary way that GI information will be communicated (because they are easier to create and print when working with stakeholders and committees), but web maps are also an option to create a more interactive experience.

However, one caution about using web maps is that they are not a *replacement* for a facilitated meeting in which the maps are reviewed and differences in interpretation can be debated in real time through give-and-take discussions. However, you can use web maps during meetings to toggle easily between datasets. The cores are symbolized using the national ranking, divided by quartile. By using only a subset of cores (those found within the county), you can show more intracounty variation of scores. Compared with the previous figure, fewer cores are shown in the following figure as part of the 4th quartile (highest ranking cores).

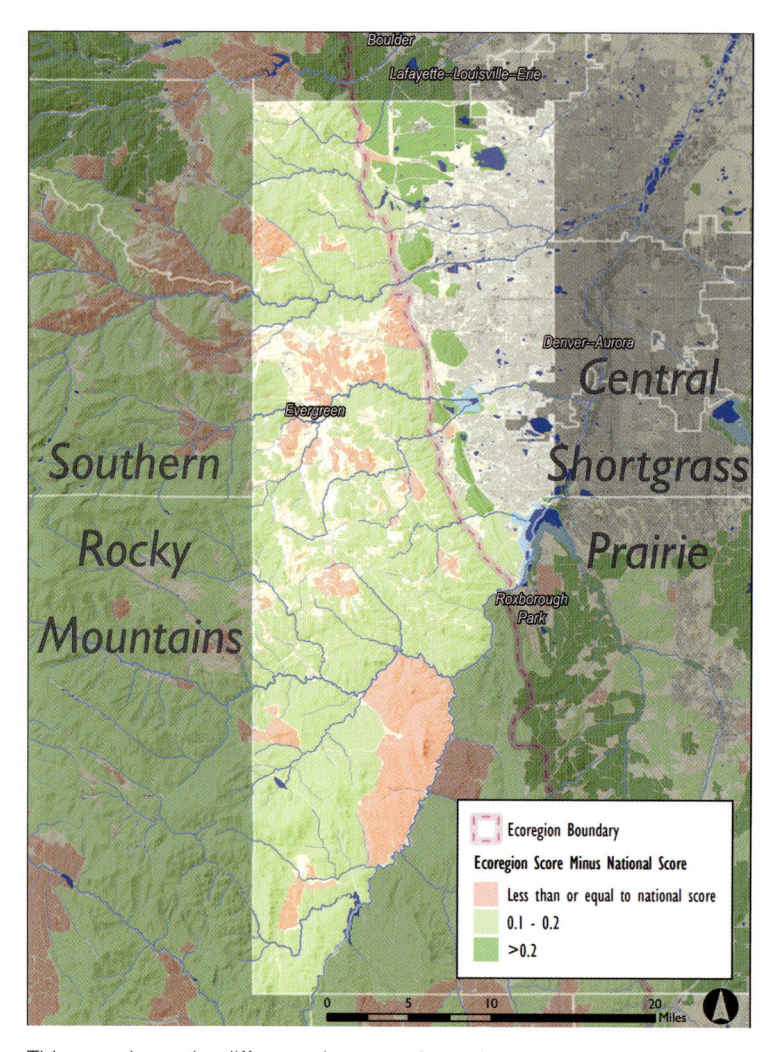

This map shows the difference between the national ranking and the ecoregion ranking in Jefferson County. The dashed pink line shows the boundary between the central shortgrass prairie ecoregion of the state's High Plains and the Rocky Mountains. Cores colored dark green have a significantly higher score when ranked relative to other cores in the same ecoregion. Cores colored red have ecoregion scores less than, or equal to, the national score. In this case, several cores do not rank highly in the national ranking but do rank significantly higher relative to other cores in the Central Shortgrass Prairie ecoregion. This observation shows why ranking cores relative to cores within an ecoregion may highlight significant cores that do not rank as highly within a national ranking scheme but are important when compared with the other cores in the ecoregion in which they are found. Credit: Esri Green Infrastructure Model.

Mapping connectivity between cores: Landscape corridors

As explained in chapter 2, a corridor is a more or less linear arrangement of habitat type or natural cover that connects cores and differs from adjacent land on either side.

Maintaining and enhancing connectivity across a landscape is a key component of GI planning, but linking habitat cores to create a network requires identification of both corridors and stepping stones, which facilitate the movement of animals, plants, and pollinators, thus preventing species from becoming isolated. Chapter 2 also raised the issue of elevation and latitude migration of large species across the landscape.

Least-cost path analysis

Least-cost path analysis determines the easiest route for the movement of species across a landscape. The analysis adds benefits and subtracts costs for using a path to determine an algorithm that provides the least costly travel route for multiple species.

In the Minnesota Central Lakes region, species might include lynx, bobcat, wolf, beaver, elk, bear, pine marten, wolverine, and puma. The region will also include fish, reptilian, amphibian, avian, and insect species. It is impossible to incorporate all the various wanderings of each species into a single pathway (corridor) that could cover the entire state. Specific corridors, or a system of corridors, must be identified that offer the fewest obstacles to these species and grants them the greatest benefits and ease of movement as they travel to the most cores possible. Identifying these corridors is a matter of identifying all the pluses and minuses—in other words, identifying the ease or obstruction of movement.

You can use the least-cost path method of raster analysis, as discussed in chapter 2, to model corridors in a GIS environment. Indeed, finding the path of least resistance is the recommended way to explore potential options when identifying corridors or maintaining connectivity across a landscape. This method is also a good starting point when considering where to preserve or enhance corridors, once you identify species needs, human pressures, and project goals.

Least-cost path analysis commonly includes an attribute that quantifies the relative cost (e.g., the PATHCOST attribute in the Connectors dataset) of traversing a path on the basis of land cover. Roads data are also useful for assessing cost and feasibility. For example, a technical advisory committee can select mapping criteria to eliminate connectors that cross major roads. What roads to include, and their relative costs to animal movement, depend on such factors as surface type, width, and traffic volume. River data and data for large bodies of water will also play a significant role in this analysis, because the data indicate total barriers to movement or a high cost to traverse.

Once you identify these least-cost pathways and potential corridors, the process must be analyzed according to the project's various goals and conflicting interests, which cannot be accounted for in a GIS-simulated reality. Therefore, a crucial step in least-cost path analysis is to assess how modeled results fit in with wider requirements, on-the-ground-realities, especially of property ownership, and other barriers that may not show up in a model.

The best way to deal with these issues is through committee oversight. This team (e.g., the technical/scientific advisory group described earlier in this chapter) can use the modeled results as one source of information, which can be combined with other variables, to understand priority connections and the real-world feasibility of maintaining them.

For example, the advisory group may use the results of the GIS least-cost path analysis, composed of hundreds or even thousands of modeled connections between habitat cores, to identify the 10 most feasible connections for preservation. The method to identify this subset of connections usually includes other key variables, such as the richness of the habitat linked by a given connection, its rarity, the risk of its being lost, the availability of alternative connections, and how actionable preservation of the connection may be. For example, does this process involve outreach to one landowner or a dozen landowners?

Tools to help inform connectivity planning

Several data products are included in the state map packages provided by Esri. A dataset of *connectors* models connectivity between cores, and a *betweenness centrality (BC) measure* estimates a core's importance in the overall network of cores and connectors.

1. **Connectors:** The Connectors dataset depicts a network of least-cost paths between neighboring habitat cores. This dataset was created using the ArcGIS Cost Connectivity tool (ArcGIS 10.4 and ArcGIS Pro 1.3). The connectors do not represent corridors but identify potential corridors. The cost surface used to create the network of connectors assumes that cost is less in an area of natural land cover and greater in areas altered by human activity.

A full description of cost values used is shown in appendix B.

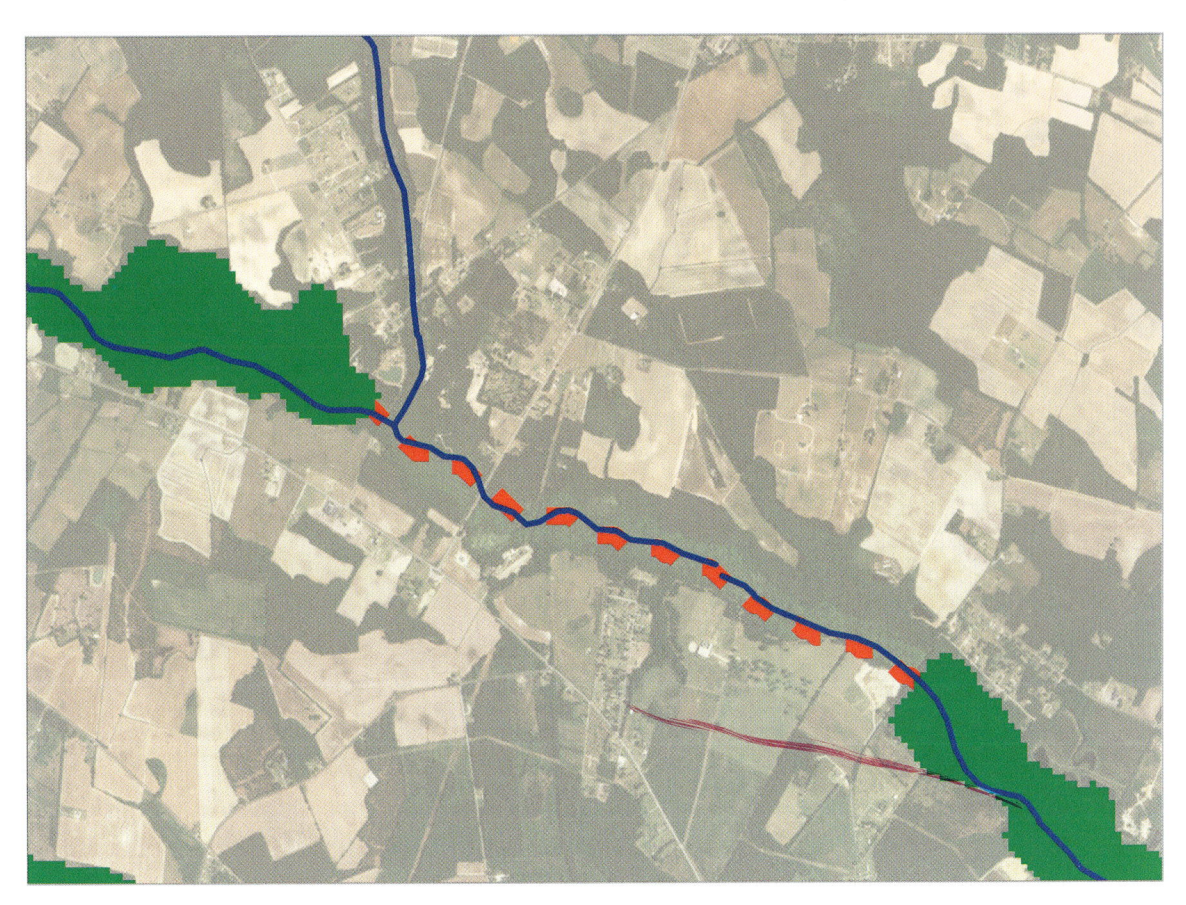

The dashed red line shows a connector that has been identified between two cores. In this case, the connector follows a stream with a relatively intact vegetated buffer. Streams often serve as natural corridors across a landscape. Credit: Green Infrastructure Center Inc.

2. **Betweenness centrality measure of cores:** Betweenness centrality (BC) provides a metric depicting each core's importance in the overall network of cores and connectors. The BC measure represents the number of connectors that flow through a given habitat core normalized by the total number of shortest paths between all pairs of nodes. Larger BC values reflect greater use of that core in traversing the network. The BC measure helps estimate the significance to the larger network if a core is lost through development or fragmented to such an extent that it no longer functions as core habitat.

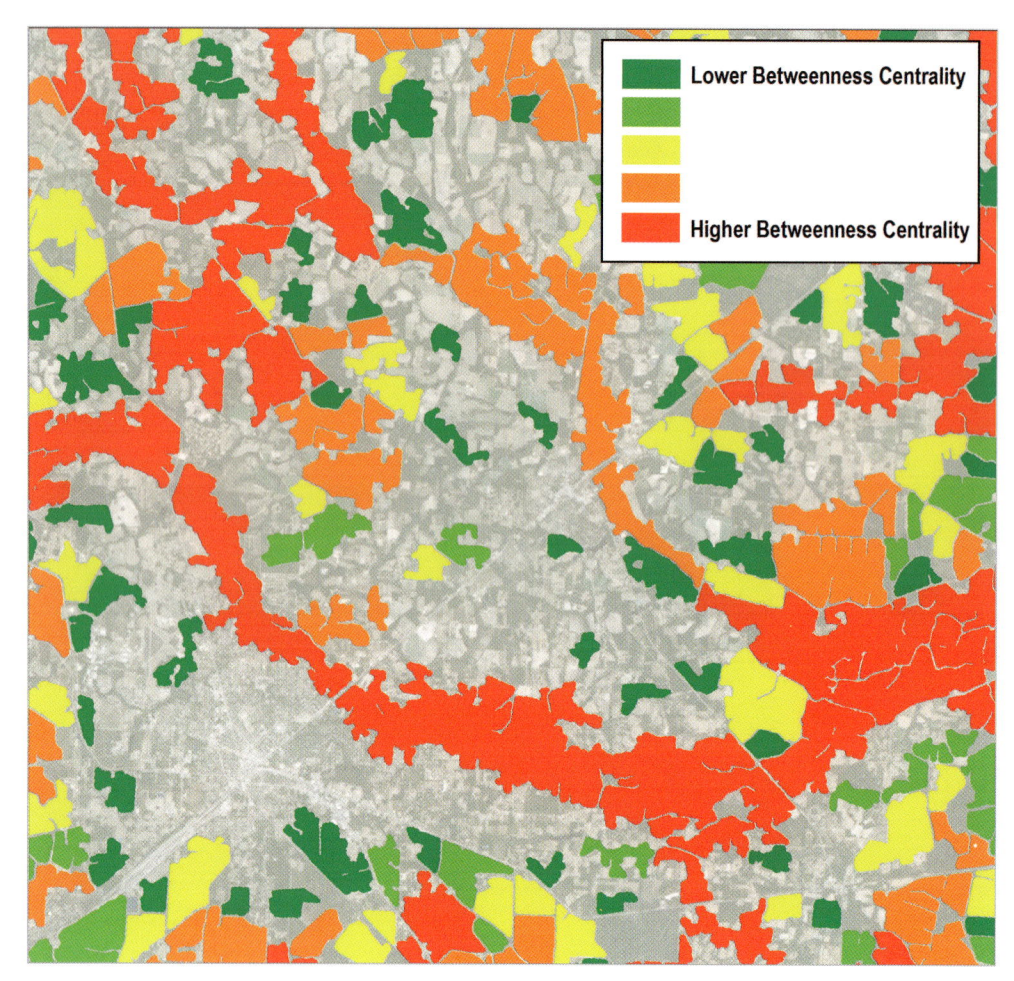

The BC measure calculated for each core reflects the importance of that core in an overall network. In this image, the cores colored red and orange have higher BC scores. Cores that are part of a connected network, such as the intact riverine corridors in this image, tend to have higher BC scores. Isolated or peripheral cores tend to have lower BC scores, as indicated by the green color on this map. A low BC score does not mean a core is unimportant, because assessing connectivity potential is only one of many lenses to prioritize core habitat. Recall that the more connected a core is, the more resilient it is, whereas the more isolated a core is from neighboring cores, the less likely it can recover from species decline by repopulation from nearby habitats. BC metrics provide insight into assessing core connectivity. Credit: Green Infrastructure Center Inc.

Factors to consider when planning for connectivity

After compiling relevant data and GIS models, you will determine the priority connections to maintain and restore. You should answer the following two questions as part of this process:

1. What are the priority connections? If you use a GIS model, the least-cost analysis likely will produce many results that will not always make sense on the ground. For example, the connectors dataset will find different least-cost paths for different species. They will

tend to ignore other variables, even if it means the paths cross an interstate highway. The project team must critically assess the results of all data and models to determine local priorities and ground-truth the paths to determine how realistic they are. In addition, such a model will not assess future plans unless the team has already mapped them. If a large highway is planned, but is not yet shown on a map, the team must account for that, either by adding the highway route, if known, to the barriers in the least-cost analysis or by hand-calculating the cores that will be lost and discarding them from their connectivity analysis.

2. What is the feasibility of maintaining (or restoring) a connection? When you take other information into account (such as future land-use plans), does maintaining a corridor make sense? A connection might have been identified as a priority but may not be feasible or expedient to maintain. In this case, another connection, not originally identified as a high priority, may become more important. Or the advisory committee weighting the feasibility of two equally ranked connections may choose to prioritize the one that affects fewer private landowners.

Determining feasibility is an iterative process that will be informed by the risk analysis discussed in chapter 6. For example, a core or connection at risk might be acquired as parkland or placed under conservation easement to prevent its future development.

Consider several questions to help prioritize connections

- Does it link large or highly ranked cores?

- What is the cost of traversing the connector? In other words, does the connector only pass through natural landscapes, or must it cross roads and other barriers or change substantially in elevation?

- Does the connector have a sufficient habitat buffer to provide the desired functions of a corridor? In other words, is it wide enough to provide the desired function?

- Can smaller cores or habitat fragments, termed stepping stones, be used to form connections between larger or high-priority cores? See chapter 2 for a discussion of the stepping stone concept.

How much of a barrier is a road, railway line, or river?

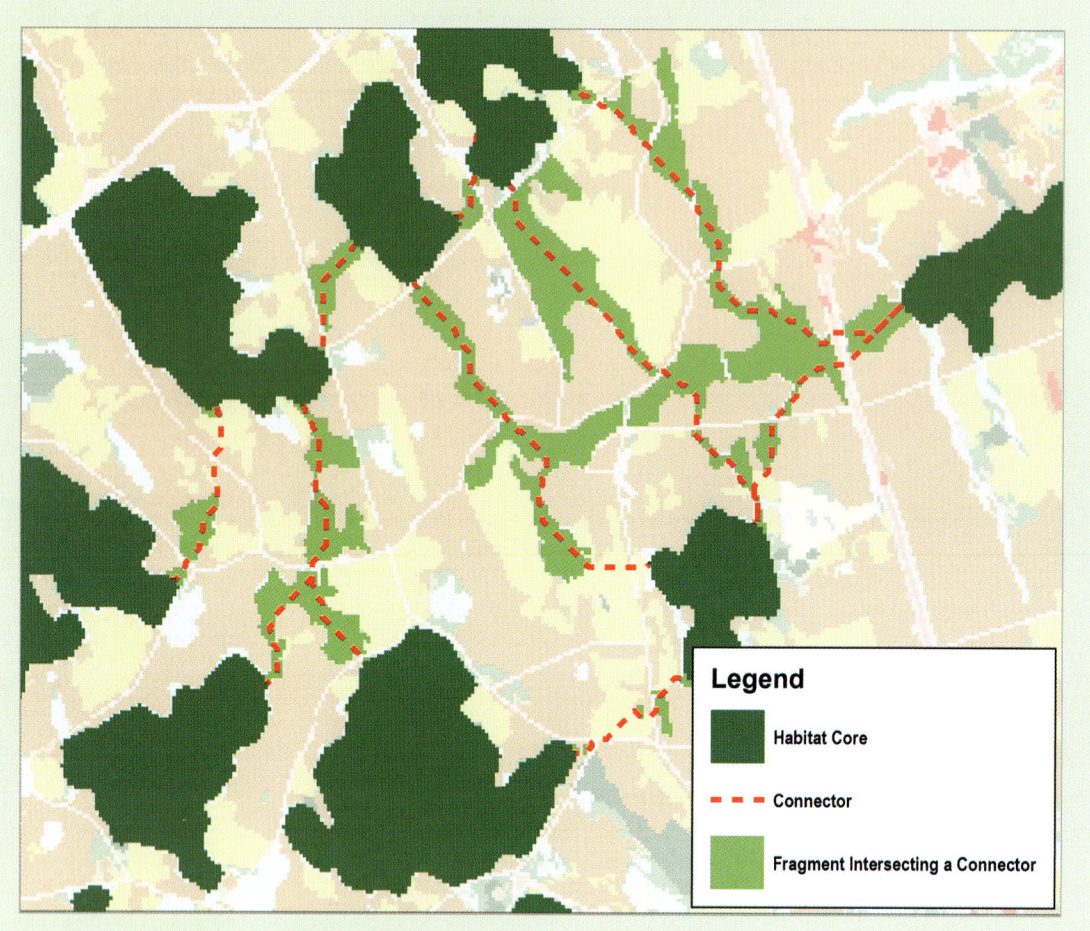

Habitat fragments can act as stepping stones, allowing species a way to move across the landscape, especially when corridors are lacking. Credit: Green Infrastructure Center Inc.

Some roads are more traversable than others. For example, most wildlife will likely cross a remote road on an isolated mountain or a dirt road of narrow width under tree canopy. In these examples, roads may not function as a bisector of a corridor, or even of a core. Generally, the wider the road (including any shoulder) and the more permanent the surface, the larger a barrier it will present, especially to smaller species, such as frogs and other amphibians.

In addition, traffic counts can be useful predictors of wildlife movement. A dirt road that connects a populated area with a town or recreation destination or that is the single entry into a state park could have a high traffic count. Railway lines also serve as predictors. A single line with little traffic may present a minimal barrier. However, a double rail line with steep embankments, fence lines, and frequent trains may present a significant barrier. You must combine width, surface, traffic counts, fences, verges, embankments, and other factors to determine whether

> ### How much of a barrier is a road, railway line, or river? (*continued*)
>
> a railway line or road counts as a *bisecting* (fragmenting) feature. You should also determine whether land cover provides a safe passage for both sides of the barrier up to the road or railway line.
>
> Include wildlife crossings in this evaluation, because they mitigate the fragmenting effects of a road or railway. A system of wildlife crossings could include underpass tunnels, viaducts, overpasses and bridges, or amphibian tunnels to reach water systems. Wildlife tunnels to link habitats have been tried with some success. Examples include tunnels under roads for Florida panthers and wildlife overpasses for large mammals over US Route 93 in Montana. Small wildlife, such as amphibians, also use tunnels to cross into breeding grounds. An example rule for mapping access would be to consider roads as barriers to wildlife passage when they are greater than 24-feet wide with more than 1,000 vehicle Average Daily Trips (ADTs). At key passage areas, reduced speed limits and signage, and even road bumps, can warn and slow traffic at highly sensitive road crossings.
>
> Rivers deserve similar consideration as possible passage barriers. Width is a significant factor, but depth and rate of water flow are more important, because many larger mammals can cross even a wide river if it is shallow and the water flow is low or is dry during some seasons.
>
> Flexibility is key in making mapping rules. Local conditions and prevalent species will dictate which actual rules are most useful. Consult with local ecologists to establish mapping rules that guide decisions on which roads and railways block animal movement.

Adding other biological information to the model

You can update the model to account for known species richness. In addition, identify each occurrence of a selected species with an *element occurrence (EO)* point or polygon to account for abundance. An EO represents locations where field biologists have observed a species or a specimen collection has been documented. The model may identify state and federally listed species and other natural heritage resources considered to have global, national, or state significance by a state's natural heritage program. Natural heritage biologists and data management staff create GIS representations of these locations, which are known as EO Reps. They map high-precision EOs with a boundary around the element location, based on survey information. They map lower-precision EOs to the best possible point based on the geographic description and buffer those points to mask the location.

Additional data concerning rare plants, fish, birds, or other unique species can be used to manually rerank a core. As noted in earlier chapters, the author recommends using data that have been consistently collected rather than data that cover only one location—perhaps a detailed study of one woodland by a local university or of a single stream by a local volunteer water-monitoring group. This avoids the problem of underranking other areas simply because they have not been monitored.

You can also create an additional overlay for sampled areas to avoid changing the core ranks when county- or area-wide species data are inconsistently available. This overlay will reveal gaps in knowledge and identify places to gather more data.

A key aspect of mapping rare, threatened, or endangered species is that data usually must be masked by placing a large buffer around any actual locations to avoid revealing details that could harm or disturb those species. Summarizing the data at the level of a whole core will mask this sensitive information. Consult the state's natural heritage program to identify what EO data exist and how to appropriately incorporate such information into the GI maps.

Creating themed overlay maps to meet goals

Themed overlay maps focus attention on a particular theme and allow the user to demonstrate how cores support a topic of interest by showing its relationship to other themed overlays and to the basemap. Common overlays include bodies of water, working landscapes, agricultural lands, recreation sites, and tourism, heritage, and culture. These themes should, in some way, relate to primary goals of the project. For example, if you developed a system of stormwater catchments and mitigation programs for an urban area, one overlay might be of impervious surfaces and another might be of ditches, drains, and streams.

The cores map primarily shows large, intact habitats, so creating a themed overlay map allows users to delve deeper into the details of a particular issue and consider actions or strategies unique to that theme. These overlay theme maps help users further evaluate which cores should be a priority—for example, overlays may reveal that one core of not highly ranked species diversity or of rare habitat type protects the public water supply, provides public recreation, and contains Native American burial mounds.

This section uses maps from a GIC-initiated project in Grayson County in Southside Virginia. At 446 square miles and a population of 16,000, Grayson County is sparsely populated and primarily rural. The county has the state's highest peak, Mount Rogers at 5,728 feet, and 35 peaks of more than 4,000 feet in elevation. The landscape features habitats that support diverse ecosystems ranging from unique bog wetlands near the Blue Ridge Parkway to high-elevation spruce/fir forests in the western mountains. Vast numbers of waterways support aquatic species that thrive in cool mountain streams. The core habitat forests support neotropical migrant birds and rare species such as the Carolina northern flying squirrel and the bog turtle that rely on Grayson County habitats for survival.

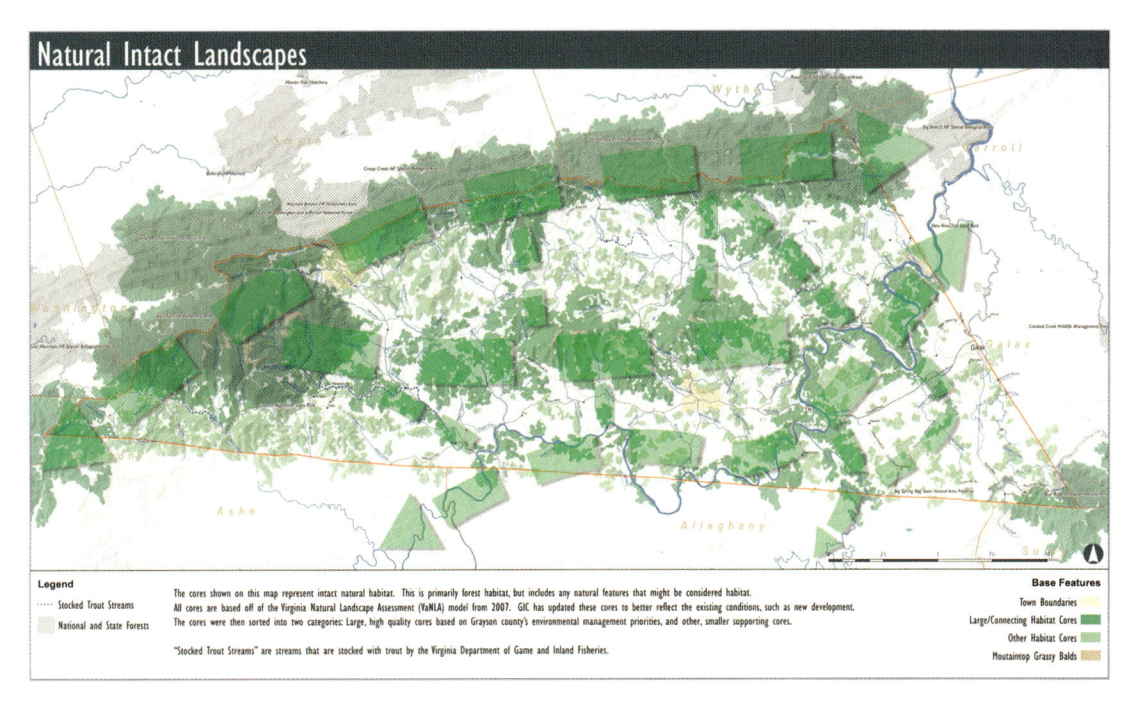

The basemap shows high-value habitat cores and major corridor flows across the county. Credit: Green Infrastructure Center Inc.

The habitat map (cores and corridors) remains the basemap, whereas themed overlay maps show the relationship of that connected habitat and landscape protection to themes of interest. Themed maps can both highlight a priority for the community and target the GI basemap toward different applications or audiences. For example, a government parks or tourism department may find the recreation map most useful, whereas a natural resource agency or economic development agency may want to utilize the working lands map. GIS users can create maps with any theme to highlight particular land uses or applications of key interest to the area.

Themed overlay maps can show new priorities. For example, if the northwest corner of a locality has a cluster of historic resources, users can prioritize the conservation of habitat cores there to conserve the landscape that supports historic structures. An overlay of historic and cultural sites may help the user upgrade and prioritize lower-ranked habitat cores to significant because they also support historic preservation. Developing themed maps can show new relationships and allow the discovery of other map applications.

Data can be used to guide action!

Water quality and watershed protection: Develop strategies for landscapes that support drinking water reservoirs, water intake zones along rivers, community wells, and extensive wetlands. Prioritize cores that support or contain these features. Use the data to guide conservation actions, such as the protection of river corridors.

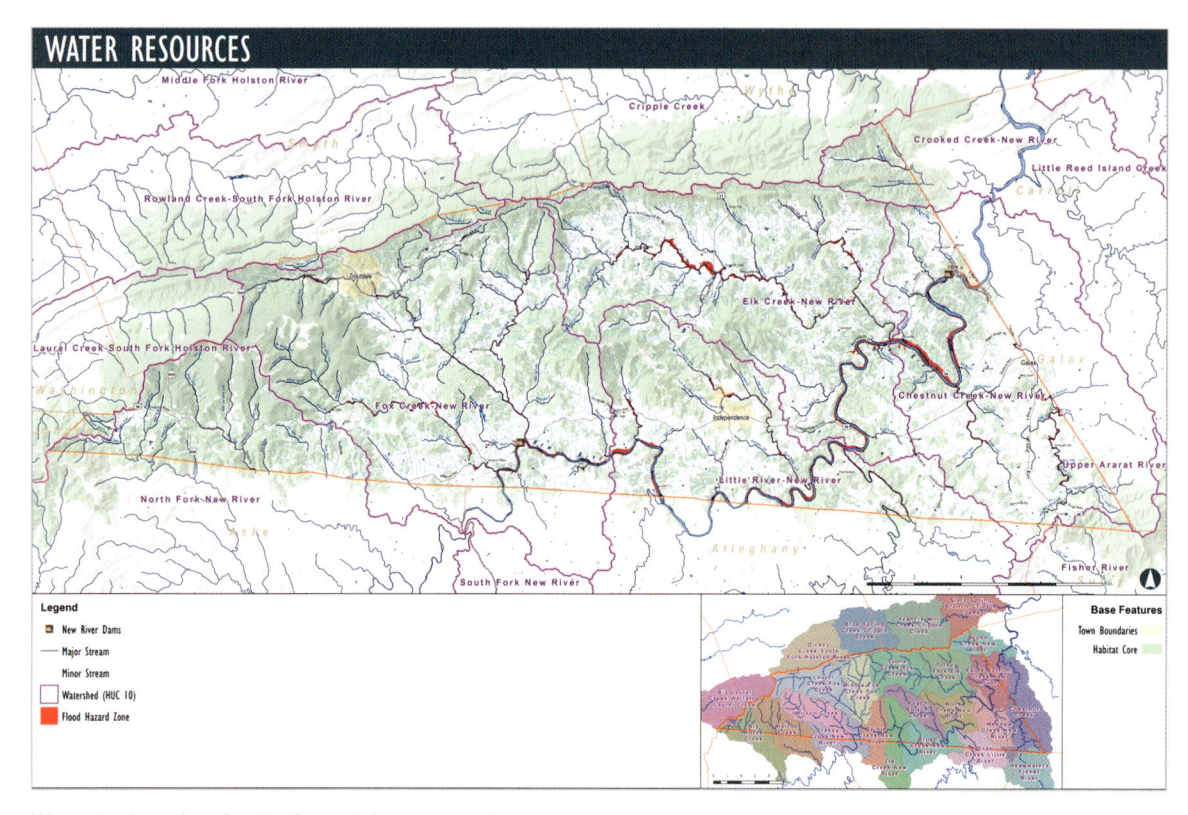

Watershed overlay. Credit: Green Infrastructure Center Inc.

Green infrastructure: Map and plan the natural world with GIS

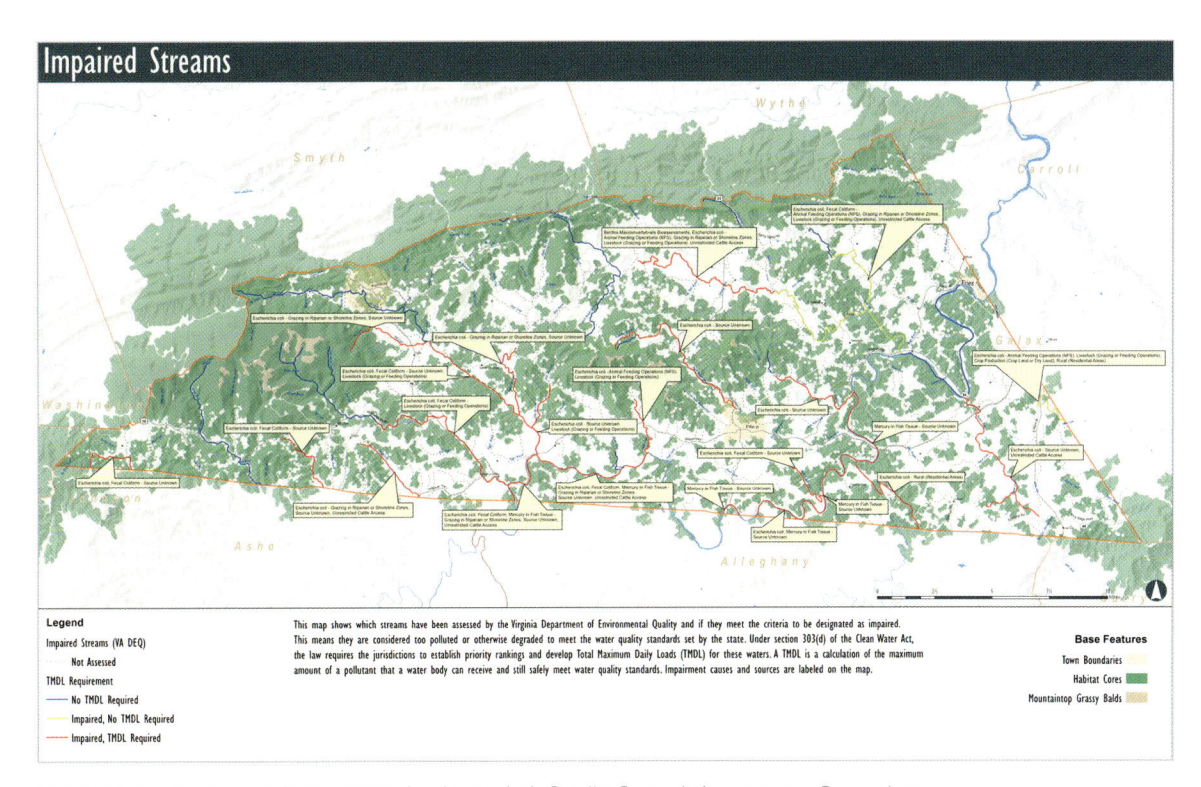

Impaired streams show where restoration is needed. Credit: Green Infrastructure Center Inc.

Chapter 5 | Making asset maps

Cultural resources: Natural landscapes provide important context and scenic qualities for cultural resources. Overlay existing historic and cultural resources to identify features either within the GI network or near it. Develop viewshed protection strategies and identify heritage tourism route opportunities.

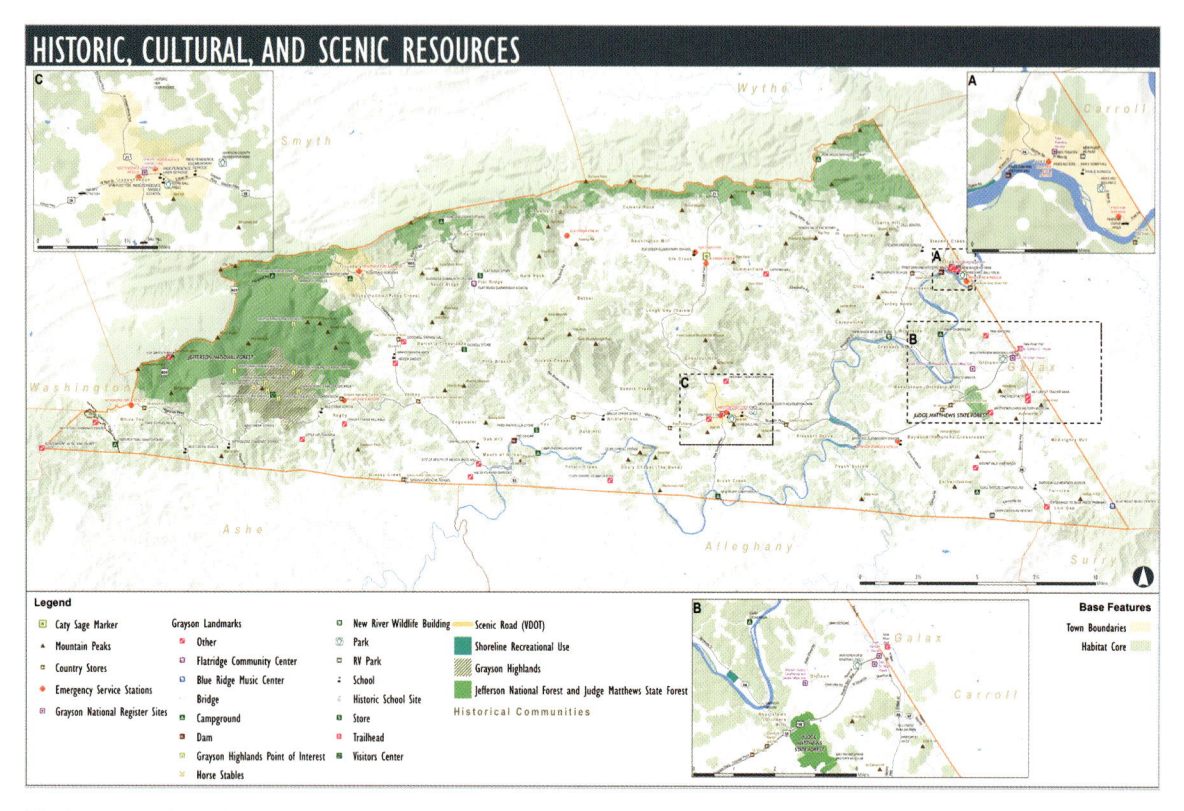

The basemap shows high-value habitat cores and major corridor flows across the county. Credit: Green Infrastructure Center Inc.

Nature-based recreation: Overlay existing nature-based recreation opportunities, such as hiking, camping, climbing, fishing, and boating, and identify key connection opportunities between cores to support trails or other uses. Maps can also inform the development of a regional recreation master plan that provides new recreational options for a community.

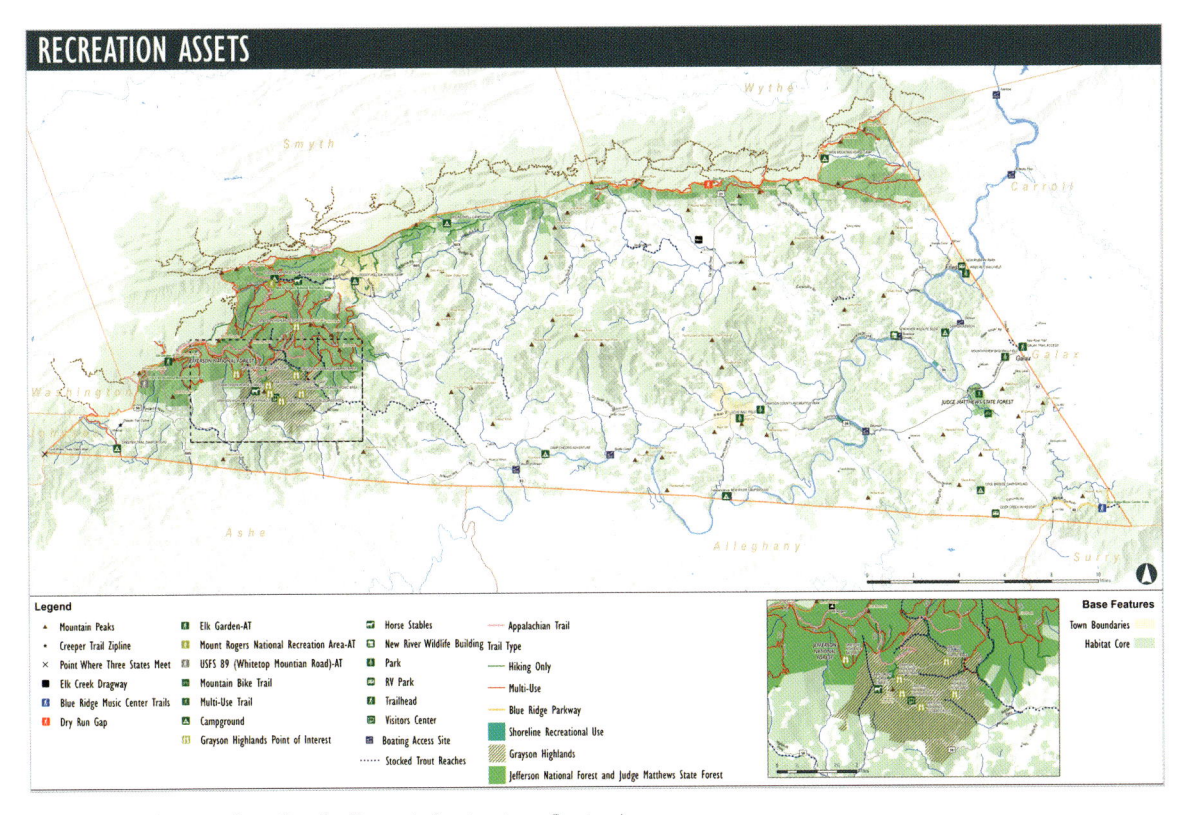

Nature-based recreation. Credit: Green Infrastructure Center Inc.

Chapter 5 | Making asset maps

Themed overlay maps: What to include

Water (elements and uses)

- **Major rivers and tributaries:** If the area has an extensive river network, consider not showing every small tributary in water-rich areas to avoid obscuring the map.

- **Large floodplains** and floodways

- **Wetlands:** Wetlands are included in the cores but can be highlighted; the map can also show wetlands outside the core habitats. The National Wetlands Inventory will not show all wetlands, especially forested wetlands or smaller wetlands. The map may also give false positives, such as standing water in quarries or farm ponds.

- **Water infrastructure:** This category includes drinking water supply intakes, community wells, reservoirs, and water recharge areas, if known.

Outdoor (natural landscape-based) recreation

- **Parks and recreation areas,** and wildlife management areas

- **Trails,** such as greenways and hiking paths

- **Outdoor recreation features:** Decisions to include these features depend on the map scale but can include such assets as fishing piers, boat launches, campgrounds, hang gliding launches, and so on.

Culture or heritage

- **National, state, and local heritage sites** and buildings

- **Known local sites:** Consider asking the local historical society what to include so you don't miss unmapped sites. If you use this approach, develop a consistent protocol for deciding what to include.

- **Sites of significance** to the community's history, such as a first settlement, fort, cemetery, or landing; campsites of colonists or explorers; and indigenous fishing encampments

- **Special viewsheds** from historic sites or that are valuable to tourism

Working lands (agriculture and forestry) that are or could be used to support the rural economy

- Highlight soils (SSURGO) that support crops; choose the highest ranked.

- Select forested areas, either inside or outside cores that are of a minimum size. These selections depend on having sizes large enough for long-term forestry management. Consult with local foresters for advice. Consider using parcel data to show minimally sized areas within a particular ownership.

- Consider working lands infrastructure related to rural lands economies, such as mills, farming cooperatives, farm markets, and other features.

- Consider other zoning or management overlays, such as agricultural and forestal districts, which can show where farming districts have been established.

PRIME AGRICULTURAL SOILS

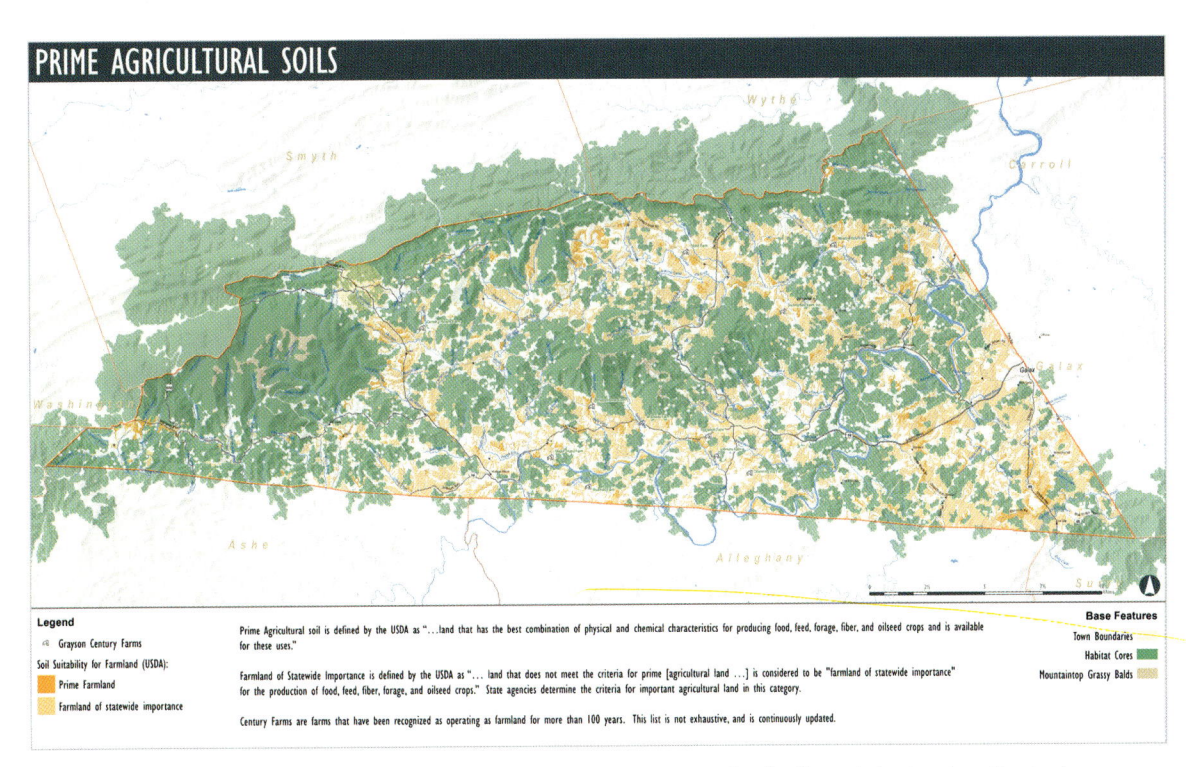

Prime agriculture soils from SSURGO are shown outside the core areas. Credit: Green Infrastructure Center Inc.

A note about working lands; the ecological core model data include only intact natural landscapes, typically forest land and other natural land covers greater than 10 acres. Using the core model to select the highest-ranked habitats as a base for forest land cover may exclude smaller forest fragments, which may have greater value at a local scale. In addition to selecting farming and forestry lands that have greatest benefits because of their size, consider evaluating smaller parcels that might support food and forestry production.

Working with SSURGO soils data: The Esri SSURGO downloaded web app

You can use a wealth of information from the SSURGO database in a GI mapping effort, but the complex nature of soils data is often overwhelming. Even seemingly simple tasks, such as overlaying prime farmland on a map of habitat cores, can confuse experienced GIS users.

Esri's SSURGO Downloader web app will help you quickly use SSURGO data. Many of the most commonly used attributes (such as hydrologic group, farmland class, and erosion class) have been joined to the SSURGO soils polygon feature class, allowing users to skip this step. You can download map packages with pre-symbolized layers by watershed using the map viewer at http://www.arcgis.com/home/item.html?id=cdc49bd63ea54dd2977f3f2853e07fff.

Forestry

One landowner will often own many adjacent parcels, whereas a huge holding often is divided into many smaller, adjacent tracts. You can obtain parcel data from the assessor's office to account for discrepancies between property ownership data and actual parcels. Then you can sort the data by landowner name and dissolve (combine) adjacent parcels under single ownership into one. This process shows total land owned by one person or corporation on that block.

This approach can cause problems, because a landowner's name may appear differently on an attribute table (e.g., Julie K. Smith, Julie Smith, or J. K. Smith), so you should check the list for similar names that may be for the same person. Then re-sort parcel data using the following categories (consult with local foresters for specific parcel sizes, because they can vary by county):

- **Twenty-five acres**: likely too small for sustainable timber management
- **Twenty-five to one hundred acres**: potential for sustainable timber management
- **One hundred acres**: suitable for sustainable timber and wildlife management

You can intersect parcels that meet the desired size with resource-quality data (such as productive soils) to identify priority areas for management or conservation. Consider adding identified areas that fall outside the GI network to the core network or represent them as contributing landscapes that meet other goals, such as promoting recreation.

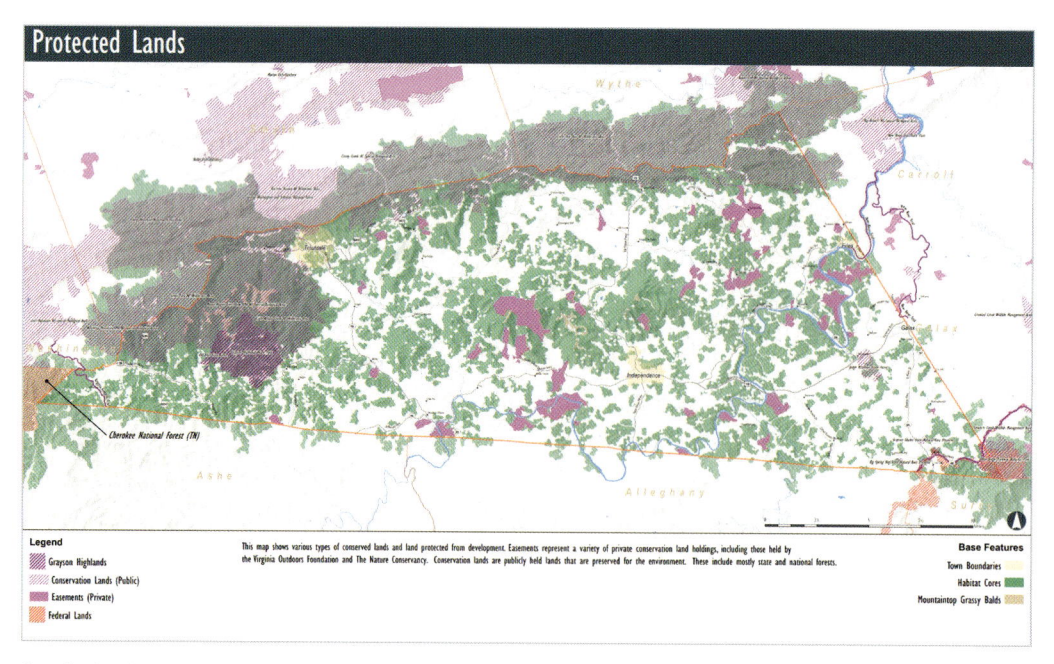

Overlaying lands protected by easement or ownership shows which high-value cores are protected and which need more attention. Credit: Green Infrastructure Center Inc.

An optional approach to refine suitability for the working lands theme

As GIS lead, you can use an alternate method to refine land-use themes:

- **Landscape suitability:** To modify the resource management layer (forest land or soil productivity), select and remove areas containing land features that make the land less suitable for agriculture or forestry management. Those features could include streams, wetlands, and steep slopes. Consult with a local forester to determine which slopes would be considered too steep to harvest sustainably without causing erosion.

Make your maps as realistic as possible. For example, although helicopter logging could technically be done to avoid slopes, it is an expensive and unlikely practice—so remove areas that are exceptionally difficult to access by selecting and removing extremely steep slopes. You can also remove unlikely harvest areas near important historic sites, riparian corridors, and identified viewsheds.

- **Code suitability:** Overlay local zoning to determine whether it is incompatible with forest harvesting or long-term forest management. Review what the zoning class allows. Is the area zoned for agriculture? Do any regulations prohibit logging? Determine whether adjacent land uses make forestry difficult and unlikely choices as parcels, such as an adjacent residential development on 2-acre lots.

- **Management opportunity:** Cross-reference working forests identified in one of the previous methods with other data on working forest lands. For example, in the US, the Southern Forest Landscape Assessment (SFLA) identifies areas with good potential for forest management. The Southern Group of State Foresters developed SFLA to highlight high-value landscapes in the southern region and target rural forestry assistance programs and funds. The project serves as the assessment component of the Forest Stewardship Program's Spatial Analysis Project. For more information about the regional map, see **https://www.fs.fed.us/na/sap/**.

Five cartography tips for GI maps

1. **Use consistent colors and symbols across maps:** This rule may sound simple but can present a challenge for a series of GI overlays. As the many maps produced during a GI plan relate strongly to each other, keep colors consistent across them. For example, themed overlays often show the spatial relationship of a certain type of asset to base habitat cores. Keeping the color scheme of those habitat cores consistent across maps is a simple way to allow readers to interpret those maps.

    GIS leads should consider features that will be shown across multiple maps. For example, showing public parks on many maps is useful because they serve as reference features and an indication of protection status. If parks are shown across multiple maps, the colors chosen to represent park boundaries must be carefully considered and used consistently, as they will be shown on theme maps with many different colors and symbols.

To quickly check whether maps have color clashes, create a map with all the layers from all the various themed maps to see how the colors and symbols look alongside one another.

2. **Avoid unnecessary detail:** Include only as much information as you need to achieve the map's intended purpose. This advice is doubly true for GI maps, because they usually present multiple types of spatially complex information—sometimes on the same map. For example, does your map need to include low-level roads, street names, or all power lines?

   On a typical county-scale map, removing minor features can declutter the map, allowing complex habitat data to be more readable. Decluttering also means reducing the number of colors or types of boundaries on the map. For example, consider what crop data should be shown on an agriculturally themed map: Do you need to show every crop type with unique colors or only the major crops produced in the county? Could you use Other as a category to simplify the map? Are parcel boundaries essential for a map of land cover types? This is not usually the case.

3. **Represent data effectively:** Think about the difference between raw data and how you might represent these data to communicate effectively to the audience. For example, represent polygon features as point symbols if they are small and hard to read, or create a point density raster, or use marker clustering if the dataset is too dense to distinguish individual points.

4. **Use the most appropriate classification for quantitative data:** Common types of classification schemes for quantitative GIS data include natural breaks, equal interval, manual interval, quantile, and standard deviation. The classification scheme chosen can significantly impact the map and how it is interpreted. Quantile classification schemes typically work well for symbolizing habitat cores data, especially for unit-less metrics such as a summary score.

5. **Show topography to enhance the maps:** In places with significant topographic features, topography can significantly help readers interpret the landscape. For example, they can distinguish cores that are mountainous and identify deforested slopes or areas where development is too hazardous. On GI maps, using a digital elevation model (DEM) to create a hillshade raster is the best way to show topography.

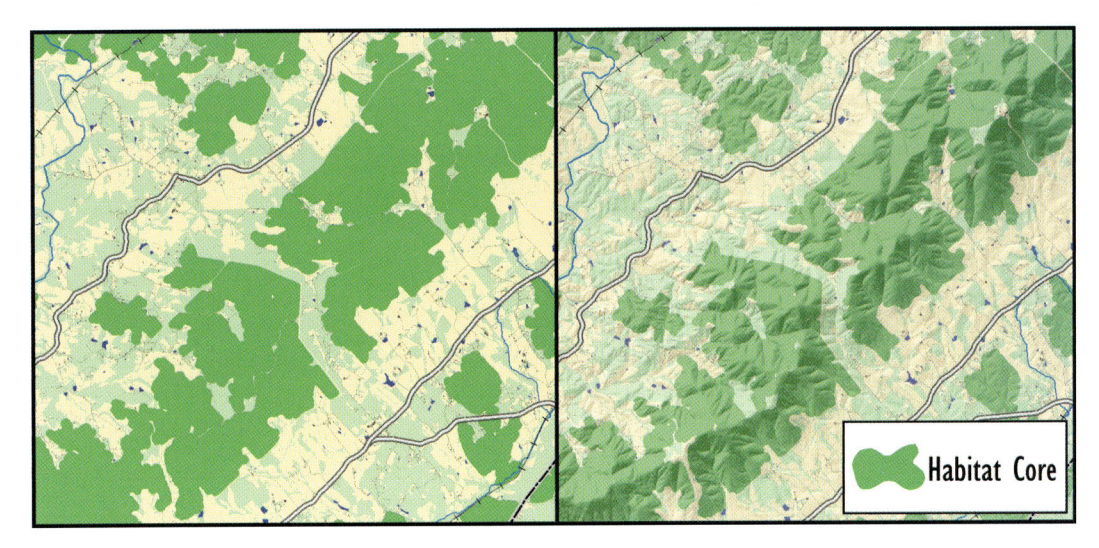

Hillshade example. Credit: Green Infrastructure Center Inc.

However, in some cases, the hilllshade layer may look too complex, perhaps because it shows minor topographical characteristics that are better left off the map. In these cases, resample the source DEM to a lower resolution—say, from 30-meter resolution to 90-meter resolution—and then recreate the hillshade. Alternatively, if the hillshade is so dark that it obscures some of the main features on the map, experiment with brightness, contrast, and transparency when you set properties in the hillshade layer dialog. If in doubt, create a subtler hillshade rather than a bolder one.

You have two options for placing hillshade in the order of layers. You can put the hillshade layer below the other layers and keep it visible by making the other layers transparent, or you can put the hillside on top as a semitransparent layer. Experiment to see which method works best for the maps.

Reranking cores to meet new priorities from themed maps

As you develop theme maps, you can overlay them on the base cores map to inform prioritization. As noted earlier, a core may be considered more important if it is found to contain rare artifacts and archaeological sites or supports community uses such as recreation or other project goals. Some cores may also become more important because they provide connectivity between two other cores or buffer a highly significant core.

Community values should determine the questions for reranking cores to meet social needs. For example, you could use a community goal to protect headwater streams to rerank all cores that are within the watershed of the water supply reservoir.

If these types of values are unknown, then the mapping process will need to include community meetings to derive more values for reranking. Note that most community members and local government staff will not be conservation planners or cartographers, so providing ready-made criteria can support their decision making. The examples that follow explain how you might use criteria based on how themed overlay maps inform core prioritization.

The following list suggests ways to prioritize cores based on topical goals. For example, if water is prioritized as an important resource to protect, you might prioritize cores within 500 feet of a stream. If agriculture is an important land use for protection, you can prioritize cores that are part of a key agricultural district, and so on. The list shows how you can translate a prioritization for a key goal, such as "Protect water quality of all surface waters" into a mapping rule.

- **Water:** Cores that
    - Are within 500 feet of a stream
    - Are within the 100-year floodplain
    - Intersect community wells
- **Working lands:** Cores that
    - Are within established agricultural or forestal districts
    - Are within 30 miles of active milling operations
    - Are buffer community gardens or large farm stands
- **Recreation:** Cores that
    - Sit within parks or other recreational landscapes
    - Buffer a park or are adjacent to a park
    - Protect scenic views, as seen from recreational areas
    - Intersect or are adjacent to greenways
    - Support hunt clubs, cross-country horseback riding, running, or mountain biking routes
    - Sit within, intersect, or are adjacent to wildlife management areas
    - Support rock climbing, hang gliding, fishing, or other nature-oriented activities
    - Provide access to water sports

- **Culture:** Cores that

    - Contain significant archeology

    - Support historic sites (nationally or locally recognized)

    - Are adjacent to or within the view from scenic roads

    - Are within the viewshed of a key overlook

    - Are part of a rural or national historic district

Cores can be ranked higher for supporting cultural uses such as fishing. Credit: Green Infrastructure Center Inc.

The rural–urban interface: Cores and the developing and developed landscape

The interface between encroaching suburban areas and natural areas is prone to many hazards that you should consider when you develop a GI plan and consider where to plan new development, all of which may impact the ranking of cores and corridors.

Fragmentation of cores in the rural–urban interface

In urbanizing areas, especially near cities, wild and natural cores fragment as development encroaches. In the US, these areas are often referred to as the *wildland–urban interface*. In such areas, houses are at greater risk of catching fire, because homes are built within forest edges or adjacent to fire-prone areas. In addition, invasive species may become problematic for the interior of the core as people plant yard vegetation that easily escapes into surrounding wildlands, their pets use the core to hunt wildlife, and all-terrain vehicles create trails. Cores in these areas may become fragmented far more than the map shows as development steadily encroaches.

Counterintuitively, cores in highly fragmented areas may become *more* important as they become less common and as the species they support lose other habitat. These cores may become the only habitat left for certain species, such as salamanders or scarlet tanagers. In these cases, you may need to rank cores higher precisely *because* they are close to human habitation. When cores are at greater risk, they can be ranked for the value they provide to people as access to natural assets close to where they live.

Overlaying these planning zones may increase the significance of some cores because they are within locally significant areas. For example, the Critical Areas Ordinance in King County, Washington, protects breeding sites (called Wildlife Habitat Conservation Areas) for all wildlife species listed as conservation priorities in the county's comprehensive plan, such as the bald eagle and the northern goshawk. Thus, a smaller core containing one of these species could be reranked as more significant to meet local priorities.

Finally, you could rank a core as a lower priority because it is within an area designated for growth, although some users may find this practice controversial. Cores can coexist in growth areas as long as they are not completely isolated and connecting corridors can be maintained. But they may already be zoned for development (see chapter 6 on risk assessment), so it may not make sense to include them as high-priority cores on a final network map.

This chapter discussed how to use the Esri model to create a basemap of intact cores, plan for landscape connectivity, and consider other local data that may inform core importance and ranking. The chapter explained how and why to use themed overlay maps to gather information by topic and how to use that information to show relationships and further prioritize cores. Finally, the chapter reviewed considerations for human values, such as access to nature from urban areas, and types of landscape risk and sensitivity to disturbance that might provide additional criteria to rank cores for protection.

Chapter 6 will review risk factors that affect the cores themselves. You can use these factors to prioritize cores for protection or to determine which cores provide lasting connectivity for wildlife. Evaluation of current and future risks will refine the basemap and point to priorities and strategies for conservation.

Note

1. Jenkins, C. N., K. S. van Houtan, S. L. Pimm, & J. O. Sexton, "US protected lands mismatch biodiversity priorities." *Proceedings of the National Academy of Sciences* 112, no. 16 (2015): 5081–5086.

Chapter 6

Assessing risks to your assets

Chapter 5 discussed the importance of mapping GI assets and creating overlays to see relationships between assets and specific criteria, usually determined by project goals—for example, how habitat cores protect drinking-water supplies, or recreational uses such as trails and scenic vistas. However, making maps and discerning relationships among various uses is only the first important task in creating a GI plan. Assessing risks to those assets is critical in deciding whether to conserve, expand, or restore the GI network. This chapter describes the risks to be aware of and how to use GIS mapping to reveal those risks. Chapter 7 will discuss how to use the asset maps and risk analysis to drive strategies for conservation or restoration.

Types of risks to map and how to evaluate them

Choosing which risks to evaluate depends on prevailing local pressures that may affect habitat cores and corridors. These risks can be *natural*, such as landslides, a rise in sea level, floods, or fire-prone areas; or they can be caused by *human factors*, such as new roads, mines, or housing developments. In some cases, human-caused impacts, such as climate change, are "natural" stressors, such as increased hurricane frequency or more severe winters, which in turn lead to more localized stressors. For example, the recent warming trend in the western US has caused arid conditions that leave pine trees more susceptible to insect infestations and manzanita scrub of the coastal regions more prone to fire.[1] During the past decade in the West, tree mortality caused by pine bark beetles has increased in spruce, lodge pole, pinyon-juniper, and ponderosa forests.

In coastal areas, landscapes threatened by climate change are also subject to inundation and can suffer greater impacts from storm surges, as waters now reach farther inland than they did, even in the recent past. This threat may require localities not only to map those GI assets present today but also to create a map of areas that will be lost as a result of future inundation. Inland areas may become more important to the GI network, because they might be the last green spaces remaining in a hundred years, as coastal land transitions to marsh and, eventually, deep water areas.

These risks arise from different sources, some controllable and some not, which raises an important distinction. Not only should risks be researched, mapped, and evaluated, but also the causes of those risks need to drive those actions taken to change them or to prepare for them. For example, the increased risk of fire in the manzanita belt could be mitigated by redirecting planned developments away from it. Rezoning could mitigate this development pressure in areas that have that tool available to them. Alternatively, new awareness of GI could prompt changes to transportation plans, including abandoning road plans or changes in routes to avoid GI assets.

Evaluate risk by level of protection

A *simple overlay* is perhaps the easiest approach to mapping risk. This overlay can depict those areas of the GI network that are conserved or not and can distinguish their various levels of protection. This overlay will determine their level of risk from development generally.

A national park is likely to be more protected from development than a local park that a county board might sell. Land protected by zoning provisions, such as land zoned as agricultural, is more protected than land zoned as industrial. Private property is likely to be unprotected from development, though conservation easements offer some protection, as do limits to allowed density. Thus, the first map to make when assessing risk is of those lands that are protected from development and to what extent they are protected.

The following lists the categories of land classification that make an area developable, from least likely to somewhat likely:

1. **Federally owned lands:** National parks, national forests, national monuments, and national wildlife refuges

2. **State-owned lands:** State parks, hunting and fishing preserves (wildlife management areas), other state land

3. **Locally owned lands:** County or city parks, reservoir and wellhead protection zones, greenways, and so on

4. **Conservation easement lands:** Lands protected in perpetuity

Several other classes of government lands, such as military bases, are also protected, but governments may decide to sell or decommission them at any time. Often, military bases contain extensive tracts of undeveloped landscapes used for war games that also support many endangered species. For example, Fort Huachuca in Arizona supports the threatened Mexican spotted owl. These owls prefer areas with a complex forest structure or rocky canyons with mature or old-growth stands that are unevenly aged and multistoried and have high canopy closure. Although not all 118.7 square miles are set aside for wildlife, much of the area's landscape is either used for occasional drills or not at all, allowing it to serve as a refuge for plants, animals, and archeological artifacts.

Other protected lands can also be added to the network to show which are protected by management status or ownership, such as state parks.

Map tip: Use crosshatching to depict protected lands so that the GI network shows through. This strategy works better than simply identifying conservation easements by different color, because some lands that are in conservation easement will not be part of the GI network.

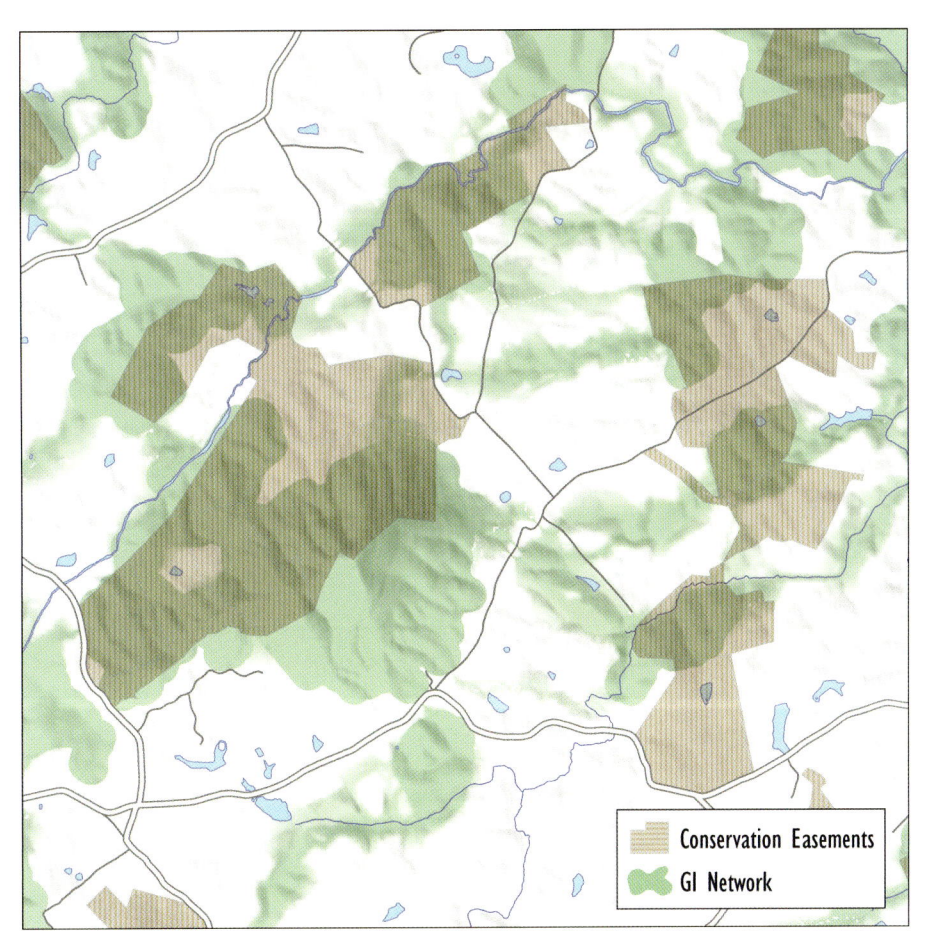

You can map conservation easements on top of a GI network to highlight opportunities for conserving or restoring critical connections across a landscape. It is often useful to show easements using hatched lines or a semitransparent fill to identify features beneath (i.e., core versus noncore habitat). Credit: Green Infrastructure Center Inc.

Chapter 6 | Assessing risks to your assets

Map tip: Consider using differently colored overlays for types of protection, such as federal or state lands, conservation easements, or types of ownership such as public or private.

Mapping the pressure to change: Evaluate threats by level of pressure and proximity of development

Private lands closest to existing developments and roads not under easement are at great risk of being developed, but that does not mean they *will* be. Other land uses near the GI network may place them at even greater development risk, such as land uses in areas zoned for development.

Also note that land protected under conservation easement may be protected for different reasons and have different value in terms of GI. For example, large ranches or row-crop agriculture may be under easement but have low ecological value when compared with other parcels. Alternatively, the protected land may harbor high-quality natural landscapes or support rare species. GIS analysts working at the local level may want to learn the reasons for an easement and code easements differently in the database depending on their intended purpose. Was the reason to protect agricultural land or a rare species? In short, mapmakers should not assume that a parcel of land under easement is of high ecological value and worthy of being part of the GI cores network, although it may meet other goals, such as conserving high-value agricultural soil for growing food.

As noted earlier, some lands may appear to have abundant habitat cores with no visible nearby threats and yet be under threat because of the phenomenon of "leapfrogging subdivisions," whereby substantial tracts of rural land are left undeveloped between housing enclaves, precisely because they thereby retain a sense of rural isolation. Alternatively, even if an area is zoned as rural or conservation, a preexisting parcel may mean the land next to it is more likely to develop. The user should determine whether that is the case. For example, can parcel owners legally build on their land if all the necessary permits in are place, such as sewage, allowed road widths, access to a primary road for ingress and egress, density permits, slope permits, and so on?

Another common difficulty is assessing the risk to large private farms that are not under easement or in agricultural zones. Such farms are often broken into large individual parcels, sometimes because the farmer's family acquired (or reacquired) the parcels over time. The family may sell part of the farm at any time, perhaps to pay off debts. Determining whether several parcels are commonly managed may require you to get the ownership records from the tax assessor's office and group them by name (keeping in mind the challenges of listings with differing spellings of the same name, such as Bill Smith, B. Johnson-Smith, William Smyth, and W. R. Smith). Remember that Bill's wife, Betsy, may have several parcels in her name too, as well as each of their children.

Chapter 7 discusses what to do about such "ghost" subdivisions. For now, you should simply flag them.

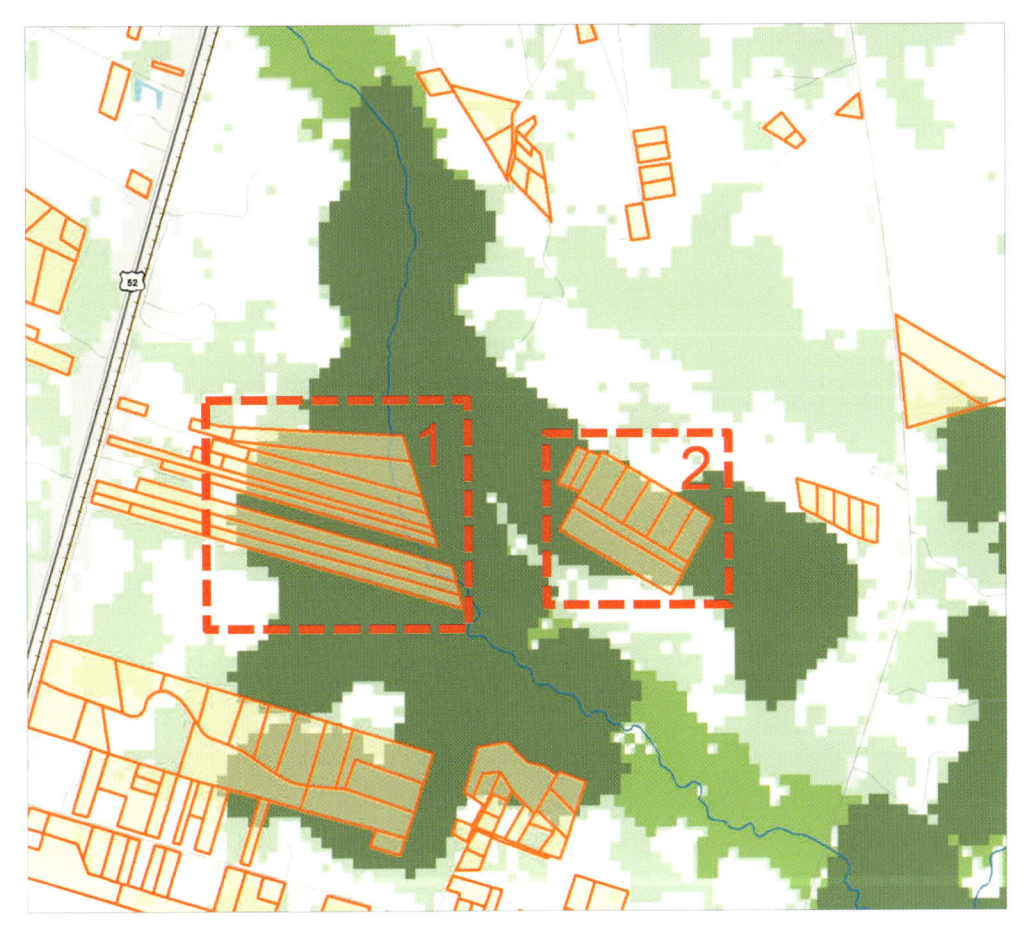

Overlaying parcels can identify threats. In this image, subdivisions by parcel 1 overlay the river, but they are narrow and may no longer be developed legally. Subdivisions in parcel 2 may be legal and could bisect the high-value core. These parcels would be a priority to acquire or to change the zoning to limit what can be built. Credit: Green Infrastructure Center Inc.

> **Examples of items that can easily be added to a risk-assessment map**
>
> - Roads, especially within 20 miles of the current suburban ring
> - Unprotected lakeside parcels
> - Areas zoned for development (residential, commercial, and industrial) or zoned as future development areas.
> - Remnant parcels in existing, densely developed areas
> - Attractions such as ski lodges, tourist spots, and national parks, which tend to attract neighbors to the area

Development causes fragmentation of the landscape, which in turn degrades functions provided by habitat cores and breaks up a connected landscape. Note the many possible risks to GI landscapes, development being one of them. Fortunately, you can use commonly available datasets to estimate development pressure on natural assets, a near-ubiquitous concern.

How to map: Performing a simple development pressure overlay

Overlay analysis is a powerful use of GIS and can assess risks during GI planning. For example, creating an overlay of factors that contribute to development pressure is a way to visualize areas that may be under more pressure to develop.

For the first step, the GIS lead will organize a team to provide advice and review. This team already may be assembled for the project or a separate team created specifically for this process. For a more complex model, it is advisable to create a dedicated team with specific knowledge about the features being mapped.

An important part of this step is to determine the amount of time or resources to dedicate to analysis. This decision will influence what type of model you use to estimate risk, whether you need more data, and what data may be obtained within the given time, resource, and budget constraints.

The second step will identify factors that influence development pressure, which you can map to varying levels of detail, depending on your time, resource, and budget constraints. You should consider the following factors, among others:

- **Geography-based regulations and policy:** You should include land-use regulations and policy in the model whenever possible as important tools for guiding and managing growth. For example, you can map zoning that encourages or discourages development. In addition, comprehensive plans may influence how rezoning, special use permits, and variances may be decided. Finally, easements and other conservation restrictions will affect land-use options.

- **The presence or absence of infrastructure:** Both can affect potential for development. Areas not served by a municipal water and sewer system will have severe limits on the amount of density they can support. Areas with roads, on the other hand, encourage development. Other factors include infrastructure plans in a capital improvement program and plans submitted to a planning department for new schools, factories, and business parks.

- **Physical features on the landscape:** Physical features can make a site more or less amenable to development. For example, steep slopes or wetlands may make development more expensive or even impossible. Flood plains are not generally good development sites, partly because of costly property insurance in those areas. Landslide zones, or zones of unstable bedrock, are a factor in some regions, notably in coastal California. Rising sea levels and depletion of the freshwater aquifer restrict further development on some barrier islands in Georgia. Similarly, a dramatic fall in the Ogallala Aquifer under the Great Plains has severely curtailed the prospects of new development in rural northern Texas, western Oklahoma, and eastern Colorado. In New Jersey, building on the beach dunes is difficult because of Hurricane Sandy's devastation in 2012.

Another spatial feature that may predict future development is lot size, with smaller lots generally suggesting more likelihood of development. Users can do a simple spatial query to find small parcels (fewer than 5 acres) that do not have any buildings on them.

In the third step, you can create a "heat map" to show gradations of risk from most likely to least likely, with the most likely colored red, the least likely green. This map can be a simple way to visualize which cores have more or less development pressure.

Limitations of risk evaluation

During risk assessment, the limitations of this analysis may become conspicuous. Many factors influence risk (including development pressure, as in this example) that you cannot easily map, or map at all. Risk assessment includes local factors, such as the specific parcel where a developer chooses to build or the relocation of a business that could reduce development pressure as jobs are lost and the population drops. The assessment includes more regional factors, such as macroeconomic trends that result from a new economic treaty or state initiative that affects growth either positively or negatively.

As conditions change, you should revisit the factors and any assumptions. Are events lining up with the risk analysis originally modeled? If not, why not? How can the model be improved?

Constraints of time, staff resources, and budget

Users should create a risk model that aligns with available time, staff resources, and budget. The method chosen because of these constraints will greatly shape the final output, so you should carefully document the limitations and resultant decisions in relation to the model recognized at the start of the process.

A binary approach

As these constraints affect most GI projects, perhaps the simplest overlay method is a *binary* approach. For each factor identified, the user can create a mapping rule that determines whether an area is under pressure—a simple yes or no, 1 or 0.

For example, a county has a designated growth area that encourages future housing and infrastructure investments. Areas inside the growth area would be considered under pressure, whereas areas outside would not. In a raster-based model, this could translate into areas selected by the mapping rule being given a value of 1 and areas that do not meet the criteria given a value of 0. A series of these yes or no rasters are then combined to identify areas that meet multiple criteria, which would be presumed to experience higher development pressure in the future.

You can identify habitat cores in such areas and tailor strategies in a GI plan accordingly. This approach is faster to perform and easier to explain. However, the approach oversimplifies the real world, particularly for risk factors that do not necessarily lend themselves to this binary distinction. Creating a binary raster based on a county's growth area is straightforward. But for factors such as accessibility to major roads, the binary cutoff is less clear and the risk is cumulative rather than binary. At what point is access to a major road no longer a significant contributor to development pressure? One mile away? Two? Three? The answer varies, and other factors come into play. Ultimately, the team must decide the extent to which it uses binary and other mapping rules and document its reasons.

A weighted overlay approach

A *weighted overlay* approach provides more nuance than simple binary data but requires more time investment from the project team. Instead of assigning a yes/no value to each factor based on a mapping rule, the team scales variables to a common range of values. For example, all factors can be scaled to a range of 1 through 9, with 1 being the lowest level of development pressure and 9 being the highest. The team may reserve a value of 0 for areas in which that factor does not apply or has no significant influence. Reserving this value is accomplished by using the ArcGIS *Reclassify* tool.

After all factors are scaled to this common scale, they can be combined to form a composite measure of development pressure, and factors that are determined to have more influence can be weighted higher. The ArcGIS *Weighted Overlay* tool is designed to do this task. Alternatively, you can use map algebra and create an expression such as:

> Composite_raster.tif = (factor1.tif * 0.25) + (factor2.tif * 0.25) + (factor3.tif + 0.5) in which factor1.tif, factor2.tif, and factor3.tif are scaled rasters with values ranging from 1 to 9.

All factors are summed after each one is multiplied by a specific weight. In this example, factor3.tif gets 50 percent of the weight of the Composite_raster.tif, whereas the two other factors each receive 25 percent of the weight. As the weights add up to 1, the Composite_raster.tif will be on the same scale as the input factors (i.e., the lowest possible score is 1 and the greatest is 9).

How to scale and weight each factor are key decisions that the project team will encounter during the process of creating a weighted overlay. The answers will vary by location. For example, the factors and weighting will likely be quite different in a rapidly growing suburban community than they would be in a more rural agricultural community.

You can refine a methodology in several ways using an iterative process. You can create multiple scenarios in which factors, scaling, or weighting are strategically manipulated. Comparing results can reveal consistent patterns across scenarios and areas that are especially sensitive to changes in assumptions. Another option is to perform a sensitivity analysis in which only one variable is strategically adjusted to reveal the degree to which it affects the end result.

As approaches are inherently flexible, they have been successfully used in a variety of situations and geographies. The Land Use Conflict Identification Strategy (LUCIS) (**https://www.amazon.com/Smart-Land-Use-Analysis-Conflict-Identification/dp/1589481747</hyperlink>**) is an example of a process that uses a series of raster-based suitability models to identify areas of potential land-use conflict. This strategy is encouraged reading for those wishing to explore overlay analysis in more depth.

The drivers of change are complex with respect to development. Not all roads spur growth, and even good zoning can be overturned. Moreover, not all development threats are equal. Many communities have expected a four-lane highway or a new bridge to spur development, only to learn that the new infrastructure (or "road to nowhere") was not built to meet a need. Another example is an area zoned for commercial growth or a new business park built in the middle of a rural area despite a lack of local market forces.

Assessing the level of risk from a road or a zoning class requires consultation with knowledgeable planners or economic development staff and inevitably involves some guesswork. Nevertheless, a heat map is one way to highlight areas to watch or investigate further.

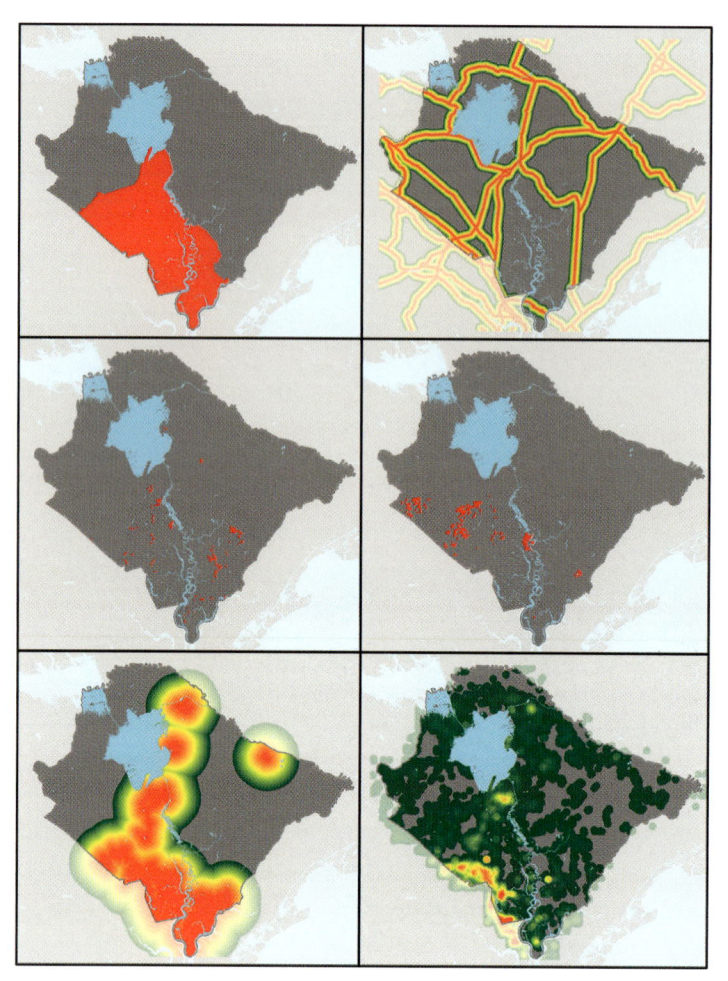

These maps show examples of several factors that may contribute to development pressure. You can map these examples and then overlay them to identify areas that may have more contributing factors for development pressure. In each image, red represents a higher development pressure created by that factor, and green represents a lower pressure. Gray areas represent areas under no pressure as a result of that factor. The image at the top left shows the county's growth allocation area in red. This example is a binary factor: either an area is in the growth area or it is not. The image at the top right shows proximity to major roads. Proximity to the risk factor affects the level of risk. The assumption is that more accessible areas will have greater development pressure, all things being equal. The middle-left and middle-right images show areas zoned for planned unit and residential development, respectively. The types of zoning districts mapped in this kind of an exercise will vary depending on local conditions (which land-use regulations encourage or discourage development).
The bottom-left image shows proximity to existing cities and towns. This example operates under the assumption that development pressure is greater closer to existing developed areas (and their associated infrastructure including schools, health facilities, etc.). The bottom-right image shows existing density of the built environment, which is similar to but distinct from the previous image. Credit: Green Infrastructure Center Inc.

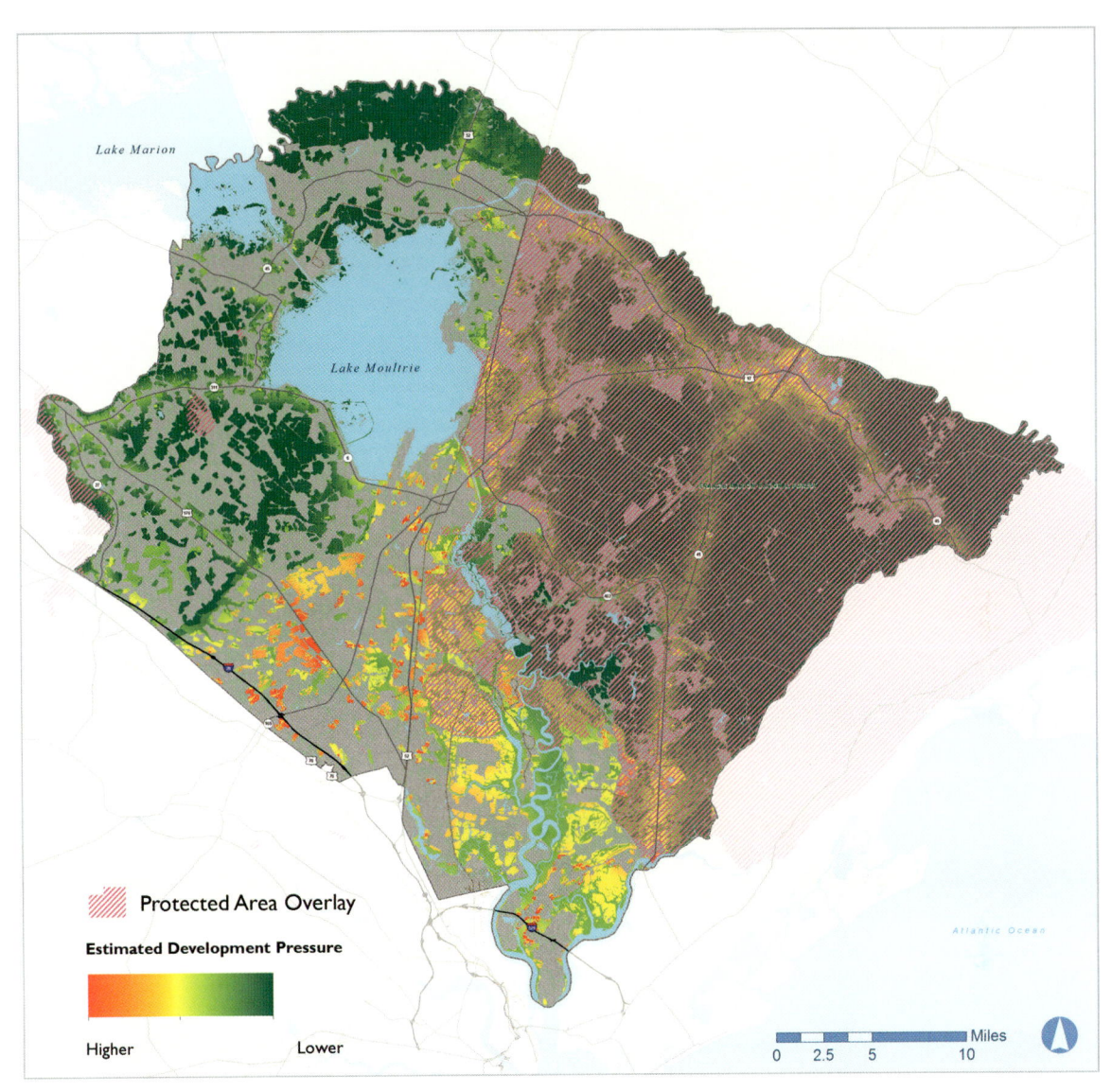

This image shows an example composite map that overlays the maps from preceding figure. Red represents greater pressure and green lower pressure, but this map accounts for multiple factors rather than a single factor. Only areas mapped as habitat cores are shown on the map to identify more easily which cores may be more at risk. Gray areas are any area not mapped as a habitat core. This map also accounts for several factors that tend to decrease development pressure, such as the presence of wetlands. For example, if an area would otherwise be shown as a high-pressure area but is a wetland, the area may be shown as moderate-to-low pressure to account for this factor. Conservation areas or easements are shown as hatched lines. You may remove these areas from the map, depending on their level of protection and the intended use of the map. Credit: Green Infrastructure Center Inc.

Evaluate by zoning or land-use designation

Overlay the zoning for an area to indicate areas at risk. Note that not all landscapes have zoning, nor is all zoning enforced or followed. Consider, too, that zoning can be outdated or even incompatible in relation to land use. Even implausible zoning uses (e.g., a remote area zoned for industrial development) are possible. Zoning simply indicates what is legally possible, not what is actually on the ground. If land containing a highly valued habitat core is zoned industrial, it can be at risk in the future.

You can combine a heat map with a zoning map to determine the likelihood of zoning becoming a reality. Or you can simply overlay the zoning on to the habitat cores to see which are zoned for development.

You will need zoning data from the local government. Choose the zoning classes (residential, commercial, and industrial) that will most likely disturb a core, whereas agricultural and forestry are the least likely. Intersect the cores with all polygons having these zoning designations. For example, rural agriculture is a compatible zoning, whereas a residential zoning district that allows houses at six units per acre is not.

Evaluate future change based on past trends

Can we use GIS to predict how land use will change, based on past patterns? The Esri Green Infrastructure web page (**https://www.esri.com/about-esri/greeninfrastructure**) features a new map for reviewing change potential developed by Clark Labs. The map allows the user to look at change predictions for the future. It uses past land development patterns and extrapolates them up to the year 2050. The tool was developed by Clark Labs, based at the Graduate School of Geography at Clark University. The tool uses an empirical model of the relationship between land cover change from 2001 to 2011 and a series of explanatory variables to predict future change. These variables include elevation; slope; proximity to primary, secondary, and local roads; and proximity to high-intensity development, open water, cropland, protected areas, and county subdivisions or counties/incorporated places (depending on the state and how zoning is regulated). Each state was modeled independently to account for differences in the intensity of change drivers, which often vary in their level of impact, depending on where they are located.

The procedure utilizes a kernel density estimation of the normalized likelihood of change associated with varying levels of each independent variable. (Kernel density estimation smooths data in cases in which inferences about the population are made based on a finite data sample.) These estimates are then aggregated by means of a locally weighted average in which the weights are based on the degree of conviction each variable has about the outcome at that specific pixel.

This tool can be an effective way for decision makers to recognize the importance of conservation planning now or expanding GI for the future. The tool assumes that past change is a predictor of future change, an assumption validated by Clark Labs developers as they created the model. The tool depicts what may happen if nothing is done to affect a different future.

However, GI planning aims to understand what landscapes are intact now and which landscapes may change in the future, and then to make decisions to alter that future, if desired. For example, will future populations use twice as much land as their predecessors (e.g., desires for 2-acre lots rather than 1-acre lots) such that a doubling of population is a quadrupling of land use? Or, could land development patterns be changed to have smaller lot sizes? Could cores be protected to steer development away from higher-value landscapes? Could transportation plans be changed to avoid sprawl-patterned development?

Enacting land conservation tools and steering growth to more appropriate areas can be difficult because not all communities have zoning. Although the market will remain the main driver in these cases, people could choose to locate nearer to higher-valued landscapes. Some developers seek plan developments adjacent to areas under conservation easement so residents can enjoy protected views that the easement affords. This example shows that the presence of a conservation tool, such as an easement, can attract new development, even though the opposite effect was intended.

Lake Tahoe as an example

Open water is considered a growth driver in the Clark Lab model, and people's affinity for water is well known. Localities often share key resources across their borders. The 191 square miles of Lake Tahoe is an example of shared management. In 1844, when American explorer John Fremont mapped the lake, his errors led to the "accidental" partitioning of the lake across state lines, with one-third in Nevada and two-thirds in California, thereafter requiring a bi-state compact for its management to balance growth and environmental protection. Even with the compact in place, a longstanding conflict has simmered between California and Nevada on how much development to permit and what to restore around the lake. According to the *Los Angeles Times*, the lake's water quality decline began when construction around the lake increased after the 1960 Winter Olympics in nearby Squaw Valley. The increase began at a time when little attention was paid to where buildings were sited, how much of the lake's nearby shore was paved over, or where sewage outlets were sited. However, a few years later, the Tahoe Regional Planning Agency became the first bi-state regional environmental planning agency in the country.[2] For more about Lake Tahoe's management and influences on the American conservation movement, see **http://www.trpa.org/tahoe-facts/history/**.

Using the Clark Lab's change calculator, at a scale of 1:288,895, the Lake Tahoe area in California has shown a 3 percent decline in forest cover since the agency was established—a loss of 9,307 acres. The Nevada side has shown a 24 percent change in developed land and a 4 percent decline in forests—a loss of 9,101 acres. The compact, therefore, appears to have kept rates of change similar on both sides of the lake, at least for now.

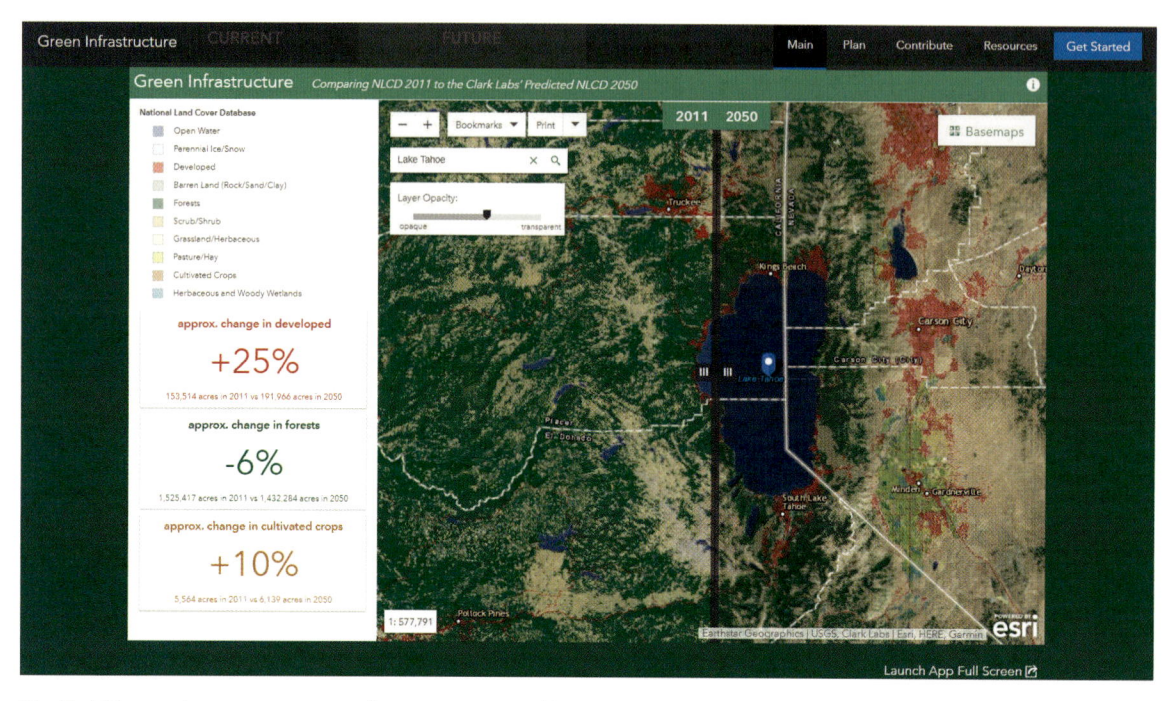

The Esri GI page features an app to allow users to consider the future and make the case for conservation action today. Credit: Clark Labs, USGS, Earthstar Geographics, and Esri.

Users of the core model should look for large cores that cross jurisdictional boundaries. Users may miss the full importance of the core unless all of it is in the focus area. As mentioned earlier, boundaries around any area of interest (e.g., a county) should be buffered to capture cores that may be large but do not appear significant when clipped to a boundary that only contains its edge. One county may have strict zoning laws, whereas another may have lax or no zoning regulations. Recognizing these potential threats and converting them to cross-boundary opportunities is discussed in greater detail in chapter 7, which discusses options for changing zoning that is incompatible with conservation of a habitat core or corridor connection.

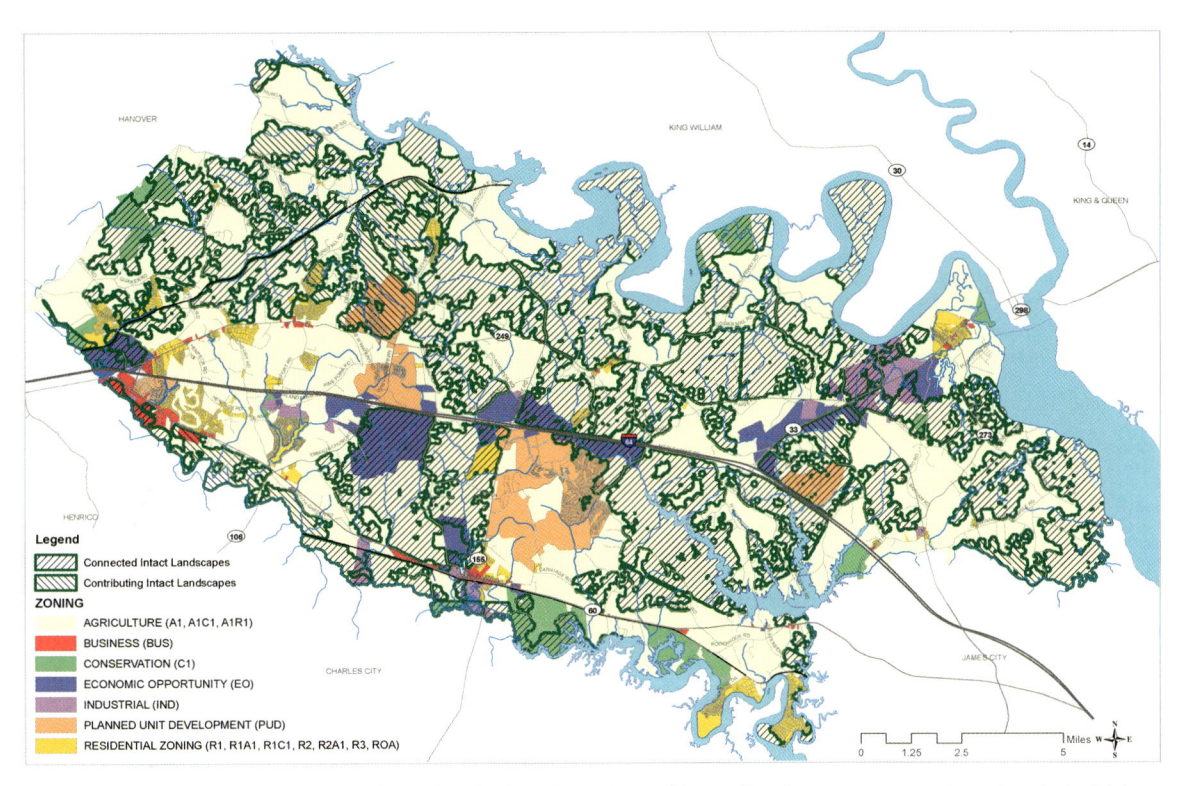

In overlaying zoning on top of cores (crosshatched to show through), conflicts become apparent, such as industrial zoning over high-value habitat cores. Credit: Green Infrastructure Center Inc.

Evaluate by known contamination

Users can also evaluate risks in terms of known contaminants that tend to pollute waterways, the air, and the earth.

Impaired (polluted) waters

In the US, the federal Clean Water Act requires that states have a program to monitor waterways and report their condition. Tribal lands also monitor water and collaborate with the US Environmental Protection Agency (EPA).

If waterways, lakes, or estuaries are found to be impaired, they may be placed on a state's impaired waters list, also known as the 303(d) list. This list can be obtained from the state agency that regulates water quality. Other countries often have similar programs. However, localities may or may not provide this information as GIS files for downloading. Contact the state or country government to determine whether this data layer is available. Determine the date of the data and whether they are still accurate, because some waters may have been cleaned up enough to allow them to be removed from the list.

You can overlay these data to determine whether the impairment may impact habitat, fisheries, recreation, or other uses. For example, based on proximity, are impaired waters a threat to drinking water or trout fishing? Understanding the magnitude of the threat will likely require consultation with knowledgeable professionals.

Impaired waters in the US usually require the development of a total maximum daily loading (TMDL) that indicates the maximum amount of pollutant that can be discharged without impairing water quality. A cleanup plan often follows. A GI plan can help meet the regulatory requirements under the TMDL requirements by conserving more land in the watershed to prevent runoff.

From a GI planning perspective, it's important to identify the type of impairment. This determination is usually noted in the data table. For example, if land runoff caused the impairment, conserving or restoring natural land cover in the watershed could ameliorate this risk. However, if the source of the impairment was an older discharge from an industrial site that is no longer present, the impairment is not related to current land cover and likely won't be affected by a GI plan. A key caveat, however, is that the impaired water could inform other goals, such as whether recreation should be discouraged in the area (depending on the level and type of threat) and whether it will impede other goals, such as expanding fishing and wildlife viewing opportunities.

When reviewing impaired waters, consider whether habitat or land-use change can address the cause of impairment. For example, if a stream suffers from excessive sediment or habitat destruction, a restoration strategy could address reforestation and enhanced stream buffers. If a cleanup plan has not yet been created, determine whether setting aside land for conservation could help restore water quality. Protecting key habitat cores for wildlife could also benefit a stream's health, depending on its location in the watershed. For more information, see chapter 7.

Once you map impaired waters as a known risk to fish and wildlife, consider whether pollution could impair more waters in the future: Are currently pristine areas zoned for more growth?

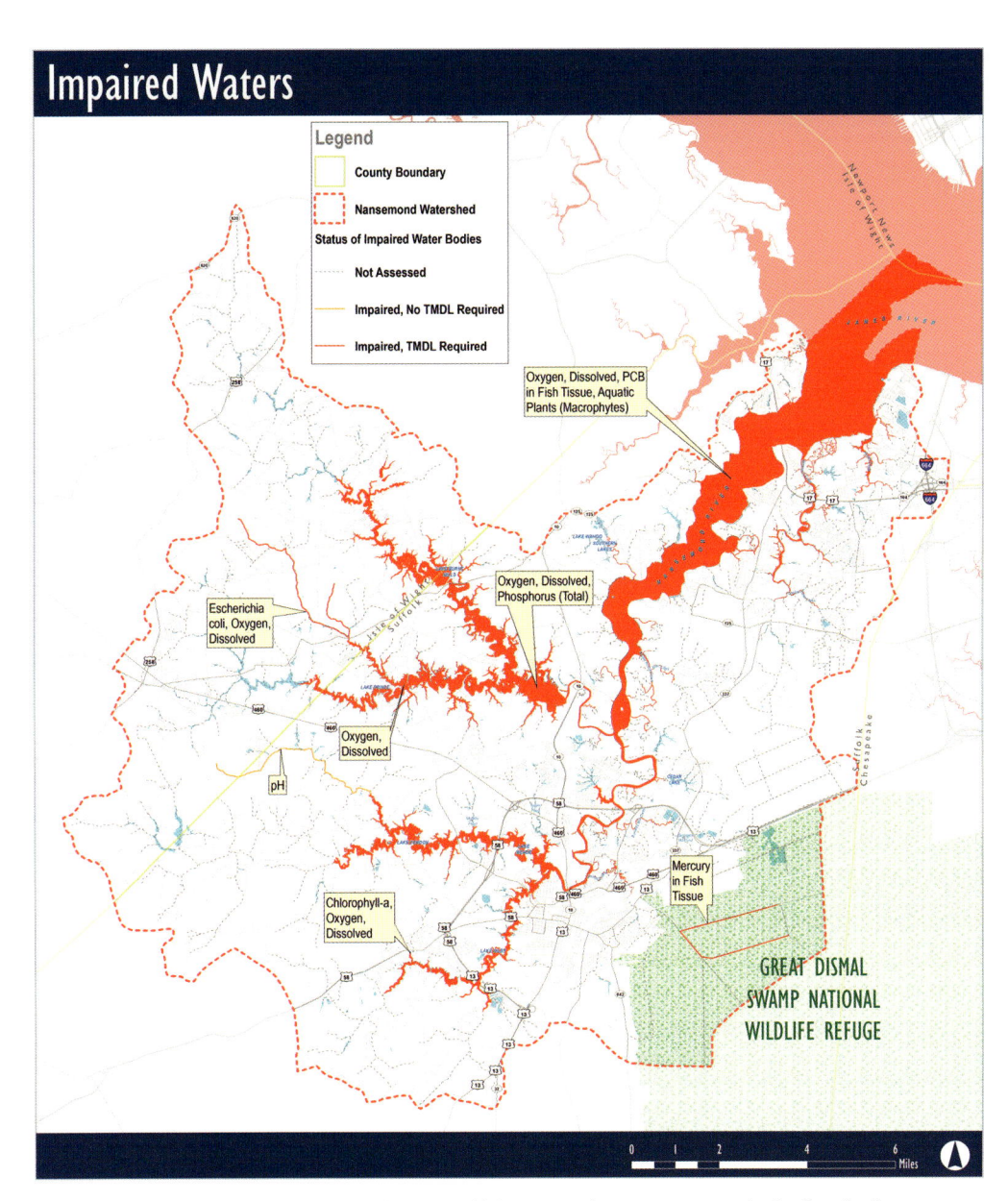

Water impairment is a key layer to add to see which waters do not meet standards. Credit: Green Infrastructure Center Inc.

Chapter 6 | Assessing risks to your assets

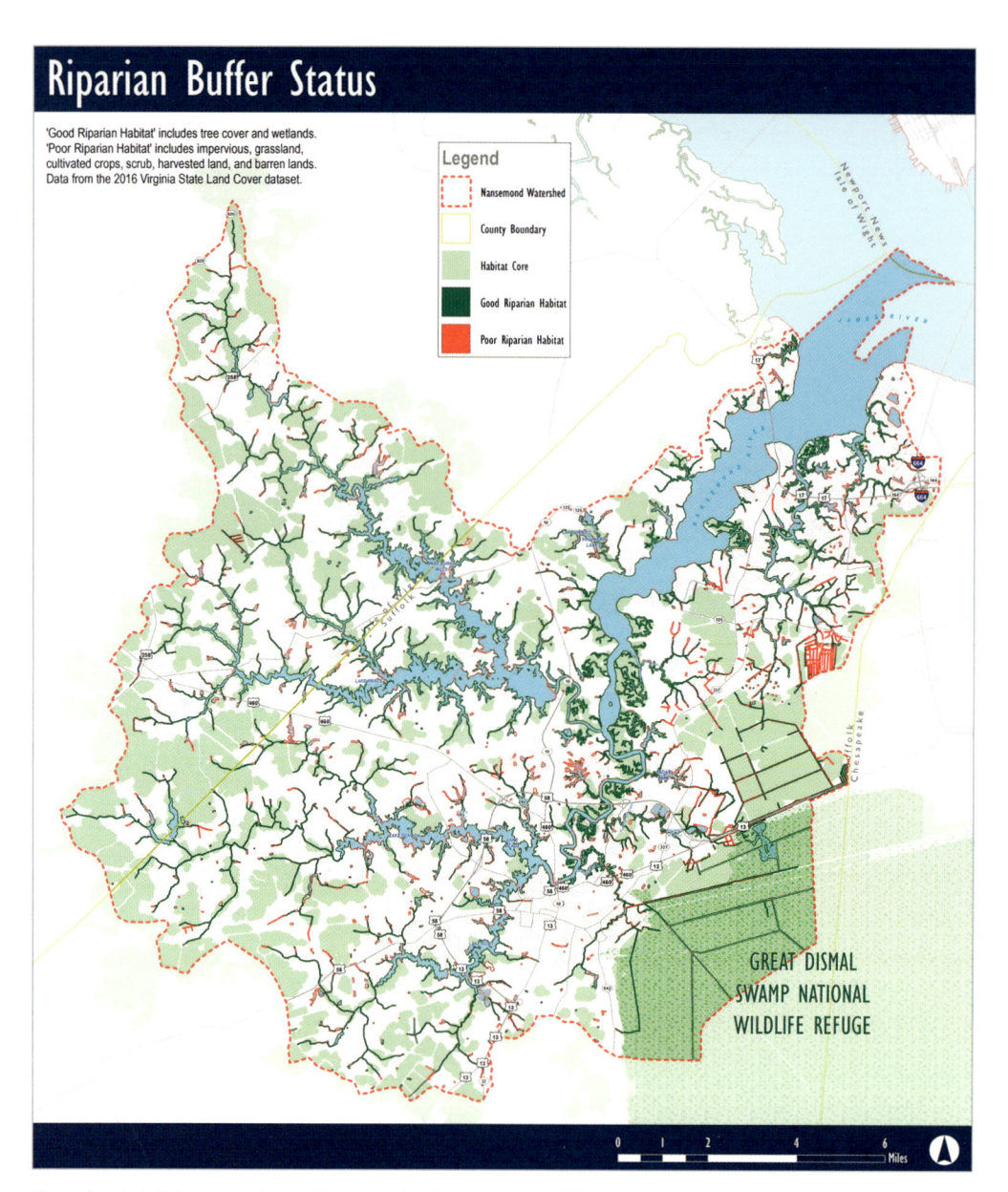

Riparian Buffer Status

'Good Riparian Habitat' includes tree cover and wetlands. 'Poor Riparian Habitat' includes impervious, grassland, cultivated crops, scrub, harvested land, and barren lands. Data from the 2016 Virginia State Land Cover dataset.

If riparian habitats are lacking, this can prioritize sites on which to plant trees or stabilize banks.
Credit: Green Infrastructure Center Inc.

The Chesapeake Bay Program

Most states do not have cross-border management compacts to manage resources. Consider the Chesapeake Bay in the mid-Atlantic region. Although it's the nation's largest estuary, the Chesapeake Bay Program that supports its management efforts is entirely voluntary. To foster faster progress, a pollution cleanup plan has established a baywide

TMDL to set limits for pollutants of nitrogen, phosphorus, and sediment entering the bay. A cleanup plan also is currently in place across a 64,000-square-mile drainage that encompasses five states and the District of Columbia. Although the TMDL across the region will reduce runoff, it may not ultimately affect land conversion, which is now the biggest threat to the health of the Chesapeake.

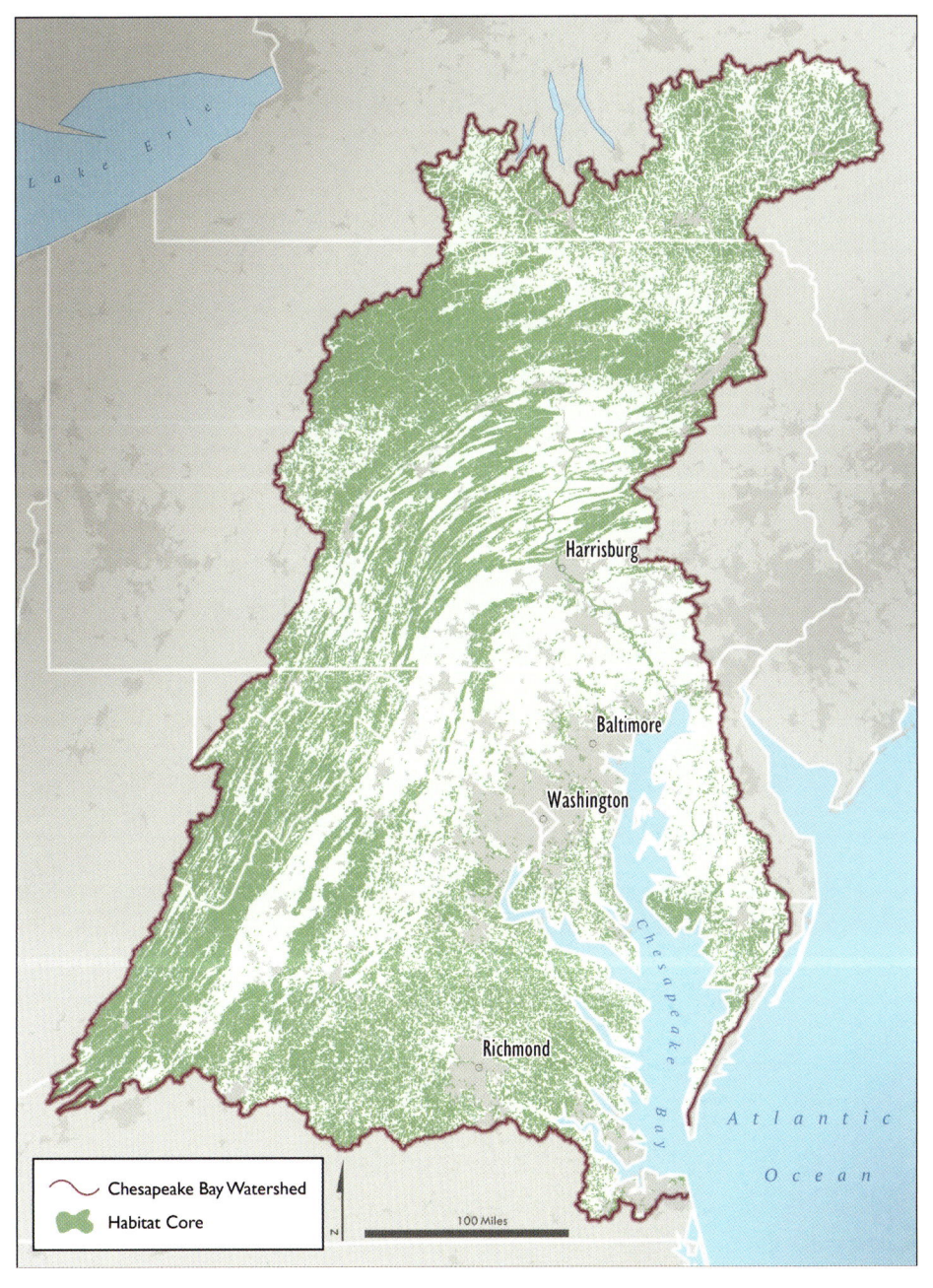

Habitat cores in the Chesapeake Bay watershed. Credit: Esri.

Chapter 6 | Assessing risks to your assets

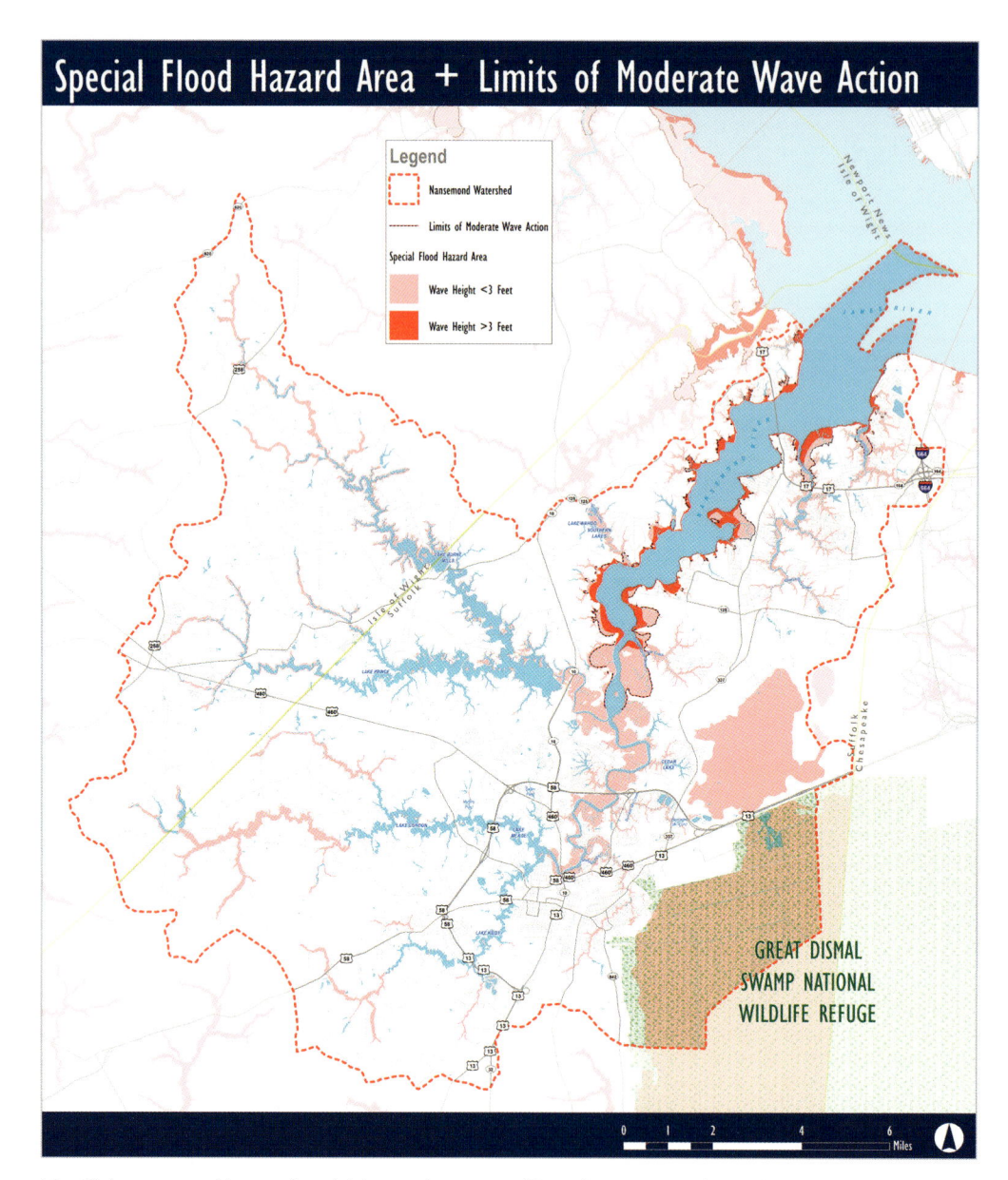

Special Flood Hazard Area + Limits of Moderate Wave Action

Identifying areas subject to flood risk can show were GI can be conserved or enhanced and development can be avoided. Credit: Green Infrastructure Center Inc.

Contaminated lands

Sites with habitat cores can still suffer from past and current contaminants. For example, within the US, many superfund (highly contaminated abandoned industrial areas) sites have become well-vegetated landscapes, fenced off for decades and left undisturbed. Similarly, dump sites in rural areas may now be overgrown but still contain contaminants. Past practices not allowed today, such as filling an earthen pit with arsenic to treat fence posts, were once common and have left their mark on our lands and waters.

You can obtain information about known contaminated sites from local government agencies and state sites that list known dump sites that have been scheduled for cleanup or capped. Some sites may need to be left undisturbed, or they may need remediation. Some sites are so toxic that wildlife—and people—should be discouraged from accessing them. You can flag these cores on the risk map for a contaminant. See chapter 7 for a discussion of strategies.

Pest outbreaks and climate change

Insects create other risks. The emerald ash borer, introduced from Asia, is currently destroying America's ash trees. Some insects native to the US have multiplied out of control. For example, bark beetles often infest dying trees and generally do not cause much destruction, but some subspecies, such as the mountain pine beetle (*Dendroctonus ponderosae*), attack and kill live trees and are thought to be spreading as a result of climate change. In 2011, they were estimated to have destroyed 41.7 million acres of US forests. For more information, see **https://www.fs.fed.us/research/invasive-species/insects/bark-beetle/**.

You can add map overlays if infestation zones are known for a pest that has caused significant damage. For example, Jefferson County, Colorado, has suffered from an infestation of mountain pine beetles. These beetles usually infest only an individual tree here and there, but every decade or so, their populations spike and cause significant damage. From the 1970s to 1980s, these beetles killed hundreds of thousands of ponderosa pines. But by 2015, the pest had subsided in most areas, largely because it had depleted the host trees and destroyed its own habitat. These infestations can be somewhat cyclical in nature. However, increasing temperatures and less rainfall have stressed those trees and made them more susceptible to damage.

Although the stricken forests may eventually recover, infestation threats should be considered when planning how much land, if any, to conserve. As habitat is put at risk, you may need to ensure that species have enough habitat to thrive. As climate change warms some areas of the earth, this warmer weather, often accompanied by dramatic changes in rainfall, can allow insects to invade areas that were once inhospitable or to overwinter in areas they previously could not, thus increasing their impacts as populations surge. Southern pine beetles are spreading farther north in the US as rising temperatures make northern climes more hospitable and insect-damaged and dying trees become more susceptible to wildfires. A recent study shows that this pest's range could reach Nova Scotia by 2020 and cover more than 270,000 square miles of forest, from the upper Midwest to Maine and deeper into Canada, by 2080.[3]

Greater fire risk and prescribed burns

You can also rank cores higher if they are fire prone. Although fire is usually considered a risk, ranking cores highly and then protecting them because of their fire risk can protect sensitive landscapes from development. Many states and regions produce fire-risk maps. You can overlay these maps on cores maps to show high-value or moderate-value cores and high fire risk. You may decide to protect these areas because they are unsound for development.

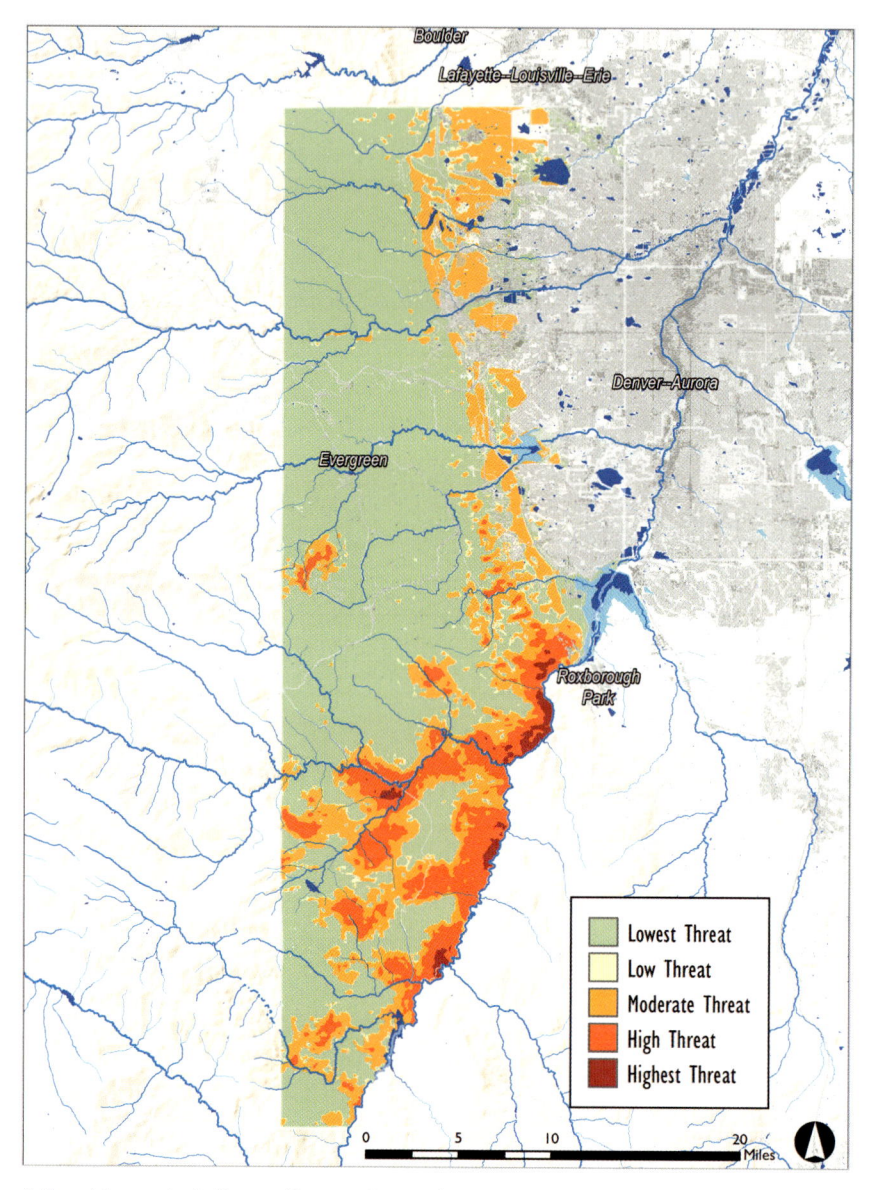

A fire-risk map in Jefferson County shows places to avoid development and to prioritize conservation. When people are already living within a fire-risk area, consider a fire-risk education program such as the International Association of Fire Chief's Ready, Set, Go! program. Credit: USDA Forest Service.

Prescribed burns are a common way to manage forests. They simulate the way natural fires help keep forests healthy and forest floor fuels, such as woody debris, from building up. However, wildfires that occur without monitoring can escape and harm people living close to wild areas. Fallen trees and woody debris are natural components of forest floors, nurturing many biota. However, years of fire suppression can leave some areas at higher risk of wildfires. Fires in these areas can burn at catastrophic temperatures, destroying not just lower trunks but entire trees. These high-temperature fires cause extreme damage and require longer times of recovery.

In some communities, smoke overlay zones discourage building, and residents must agree to periodic smoky conditions as foresters conduct burns to lessen the possibility of catastrophic wildfires. You could add these zones to a map of GI risk or simply note them as information concerning where development is more or less appropriate.

Using zoning ordinances and other protective measures to provide safety

Special zoning district overlays may signal that an area is important and worthy of further protection. Users should consult local planning agencies and governments to learn whether areas have been specially designated as locally significant. Users could then flag those cores as more important (increase their ranking). For example, Florida designates certain districts as Critical Wildlife Areas, often to protect key nesting or migratory sites. In the developed landscape of Falmouth, Massachusetts, the community created a Wildlife Overlay District in which proposed developments must take practical measures to protect habitat.

Climate change

Climate change also affects migration and general conditions for animals. Impacts to polar bears from melting ice are well publicized as they dramatically lose their ice sheets. However, less obvious impacts affect wildlife such as moose, which may be affected by higher parasite predation, as warming temperatures allow northward migration and more pests to overwinter. Tens of thousands of parasites can infect a single moose, weaken its immune system, and kill adults and calves, which are most vulnerable.[4]

Species change can also occur as a result of hydrologic changes downstream from melting glaciers. In the US–Canadian Waterton-Glacier International Peace Park, the namesake glacier is shrinking, opening up new areas for some species while disrupting habitat for others. The rapid melting can have effects on aquatic species by changing stream water volume, water temperature, and runoff timing in the higher elevations of the park, according to US Geological Survey (USGS) scientist Dr. Daniel Fagre. The food sources of a species may also change and disrupt its distribution. For example, drier weather in California has reduced the food and water supply of the desert bighorn sheep at higher elevations, contributing to their migration into valleys of lower elevation, where it is more susceptible to predation, diseases, and hunters.[5]

In the Netherlands, birds such as great tits are suffering as warm temperatures that now occur earlier in the year signal caterpillars to emerge sooner—but too early to benefit the great tit's young, which now hatch too late for the feast.[6] As species move to new habitats, they may disrupt food sources for native species as they compete for shared food resources. Although species can often adapt to changes in their environment over time, the question is whether they can adapt quickly enough to avoid extirpation.[7]

Evaluate by disrupting features

Some features that disrupt and fragment cores—such as power lines, rail lines, and large roads—are already accounted for in the model of habitat cores. However, remember that Esri constructed its model using land cover from the National Land Cover Database, which is created every 5 years, so new roads may have been constructed since then, rendering a core obsolete.

To evaluate risks, a user should distinguish between possible and probable risks. This will take further research and finesse. Many regions publish road plans, but not all those roads are funded, and some may be dropped entirely as priorities change. These risks can be classified as *possible*. In contrast, a highway project that has obtained all the necessary permits and is scheduled to begin work in 30 days is a *probable* risk, because it is extremely likely to happen. The user may want to overlay the route generally and clip out the road's right of way and a buffer to indicate disturbed areas. These risks can be accounted for more easily by meeting with a local transportation planner, which will save time researching impacts or the likelihood of a project coming to pass.

Pipelines and mines

Similarly, many of the planned pipelines and power lines in the US and across the globe will never get built. Projects that are built may cause disruptions during construction, but they will eventually become less disruptive, especially if they are buried and don't leak. Planned pipeline routes also are subject to change. Using GIS, planners have the advantage of running new analyses as routes of roads, pipes, and power lines change.

When an impact likely will happen, maps should show the route along with any land that must be kept cleared alongside it and a buffer of 300 feet on either side of that to account for the edge effects of continued disturbance. If the area to be cleared is known to be exceptionally wide, say, 1,000 feet, add that width by buffering outward 500 feet from the line feature of the pipeline. Impacts extend beyond the pipeline. Staging areas, equipment storage, construction sites, access roads for monitoring pipelines, and processing facilities can all add up to a tremendous impact.

Fracking is another disrupting feature, as is mining. Although proponents of fracking often note that pipelines have small footprints or that rail lines move their product, they may not be accounting for similar large areas used for staging areas, road networks, storage, and other needs for laying a pipeline. Similarly, large mines can permanently disrupt an area, especially if they involve mountaintop removal or open pit mining, which are increasingly common methods of extraction, because they are often less expensive than tunneling. If a mining or fracking operation has already been permitted, but not yet started, the area open to that use may be flagged on a map as an area at risk.

Less obvious risks include mineral rights, which may be recorded only on an old land deed and may be difficult, if not impossible, to uncover. For example, an area may suddenly be opened to mining in a remote area where no current mining is occurring, especially if there is a sudden world shortage of the ore.

Reclaiming old mine sites

Although mined lands often can be recovered somewhat, rubble and mining spoils fill many reclaimed areas that do not replicate their pre-mined condition. Two centuries of US coal mining occurred before the passage of the Surface Mining Control and Reclamation Act (SMCRA) of 1977, which sets some standards for post-mining site reclamation.

The Abandoned Mine Land Reclamation Program addresses the hazards and environmental degradation posed by legacy mine sites. Even in places where mined land reclamation funds are available, such as West Virginia and Kentucky, funds go untapped, even though these areas still suffer impacts from mining operations that ceased decades ago. Some states provide maps of abandoned mines and information on related hazards, such as whether a mine contains slurry piles or contaminated waters.

Natural hazards

The book has briefly touched on natural hazards in passing, but when it comes to risk assessment, more detailed analysis is needed.

Landslides and avalanche zones

Landslides and avalanche zones are areas where habitat may be at risk of frequent disruption. Natural changes to the land are worth mapping to highlight possible wildlife impacts and identify areas unsafe for recreation and habitation. Some landslide areas are well known, such as State Highway 1 through Big Sur along the Pacific coast of California.

Many avalanche areas are also well known. Climate change could increase the frequency of avalanches, as scientists have observed that changes in snowpack density (and thus snowpack stability) also increase the avalanche risk.[8] Avalanche zones can be mapped according to level of severity, as well as their potential impacts on life and property. Standards adopted from Switzerland offer guidance on how to designate avalanche zones, although there are no national US standards.

Avalanches typically start on unforested slopes or slopes with a steepness of between 30 and 50 degrees. In coastal climates, they tend to occur on steeper terrain, whereas heavily forested slopes are less likely to move. In higher latitudes, lee slopes greater than 25 degrees tend to be at risk. Areas with gullies and natural drainage channels also provide pathways for avalanches, and broad, unconfined slopes are also subject to avalanche.[9]

Natural events usually become hazards only when they disrupt people. However, they can affect wildlife too, so including wildlife on a GI map of habitats may be worthwhile, even if an area is devoid of people. States subject to avalanches often maintain avalanche data, so GIS analysts should contact their state to learn whether it has such as a data layer. For example, in Utah, the Utah Avalanche Center maintains a map database in partnership with the US Forest Service at **https://utahavalanchecenter.org**.

According to the USGS, you can map landslide dangers from historically known landslide sites that indicate the likelihood of future events. You can compile these locations into quantitative maps based on probabilities using variables such as rainfall, slope, soils, and proximity to known earthquake areas. *Landslide susceptibility* (also called *potential*) maps depict the likelihood of a landslide based solely on the intrinsic properties of a locale or site. Three of the most important site factors that determine susceptibility are prior failure, rock or soil strength, and slope steepness. If the potential for loss of life, property, or services is considered, the map is called a *landslide risk map*.[10]

Lidar makes mapping potential landslides easier. For example, researchers with USGS recently mapped many previously unrecognized landslides in heavily forested terrain in the western Columbia Gorge, Skamania County, Washington, and determined that some previously recognized areas of instability were composites of multiple smaller landslides.[11]

Sea level rise

How does sea level rise (SLR) relate to GI planning? In mapping habitat cores, you should consider mapping current habitats and projected changes to coastlines and habitats as sea levels rise. Migrating coasts tend to form habitats that occupy new areas farther inland. These habitats can impact low-lying areas, such as cities along North Carolina's coast, and border extensive natural coastlands, such as marshes.

Most of the world's scientists agree that the increasing rate of sea level rise is caused by the earth's warming atmosphere and seas and the melting of glaciers and ice caps. Most scientists agree that climate change results from human-made emissions of greenhouse gases. However, climate change is treated as a natural hazard, because the result is rising of ocean levels.

Sea level rise causes rivers, bays, and coastlines to shift inland. This rise is sometimes difficult to observe, partly because of previously inaccurate measurements of past shorelines. Coastlines experience uneven SLR partly because the ocean floor is uneven. In addition, areas with natural bluffs and dunes are less affected than low-lying areas where just a foot of rise can inundate once-dry areas for long distances inland.

Some coastal landscapes are also subsiding. In the US, the Louisiana bayous and New Orleans are subsiding because of a combination of historic water withdrawals and levees that block water from carrying sediment to replenish what is ecologically a giant alluvial fan. Starved of that sediment and having its groundwater diverted and drained, much of the City of New Orleans is now below sea level, kept dry only by a system of levees and pumps. Those systems face the threat of being overwhelmed, as they were during Hurricane Katrina, but not just temporarily. Similarly, cities such as Miami, Florida, and Norfolk, Virginia, are experiencing land subsidence simultaneously with sea level rise.

Encroaching seas will transform habitats, perhaps turning a shallow marsh into a deep-water habitat and a dry area into a marsh. Rising seas could drown a coastal forest or change species compositions from those that inhabit hummocks (raised areas of land) to those that inhabit swamplands.

Species have shown they can adapt naturally when change happens slowly. Wetter species will replace drier species, and upland species will steadily colonize upslope areas. The challenge occurs during more rapid change when a marsh or forest has nowhere to migrate, either because of a natural barrier, such as a steep cliff, or human barriers, such as dams, levees, and paved urban areas.

In the US, you can obtain SLR data from the National Oceanic and Atmospheric Administration (NOAA) at **https://coast.noaa.gov/slrdata/**.

To use SLR data, you should consider the year to which you wish to project the model. If you plan for the next 20 years, does it matter if the area will be underwater in 100 years? The time horizon for the GI project will dictate the level of SLR that is mapped. For example, when SLR was utilized for a GI plan for Norfolk, Virginia, the question of when to consider a shoreline or forested buffer as a viable project was considered. Since the plan was adopted in 2018 and taking into account the need for 2 years of project planning and funding, and a decision that a habitat project should have a 20-year window of benefits before rising waters erase the project, the year 2040 was used for the cutoff for project viability. The year 2040 was found to correlate with an SLR of between 1.5 to 2 feet, and so this SLR encroachment was mapped and formed the spatial basis for the placement of projects. Those areas shown to be underwater by the year 2040 were removed from consideration.

In other words, you face a far different time horizon if you are planning a bridge costing millions of dollars that is expected to last 75 years, compared with a time horizon for a relatively inexpensive forest buffer planting project that provides immediate benefits, even if SRL inundates the project after 20 years.

Once you determine what year to map to, you can overlay the SLR on the cores map to see which areas are at risk. First, investigate the data available for the region (the *USGS Sea Level Rise Modeling Handbook* at **https://pubs.usgs.gov/pp/1815/pp1815.pdf** is a good place to start). NOAA, the Army Corps of Engineers, and universities often have models and data available for little or no cost. Data that estimate inundation and ponding at intervals (1 foot, 2 feet, and so on) tend to be more flexible than models that estimate inundation at a certain point in time, because estimates change, but elevation remains fairly constant. As a result, you can map an inundation level consistent with a county's or a city's strategic priorities.

You may prefer to create an SLR model if one does not currently exist, or if existing models use coarser input than is currently available—for example, a highly detailed, LiDAR-derived DEM. You can produce a simple estimation ("bathtub" model) of areas at risk from SLR by creating a subset of a digital elevation model wherein cell values are less than the estimated SLR. For example, you can use the ArcGIS Raster Calculator tool to find areas of elevation less than 1 meter above sea level by using an expression such as "DEM" < 1. A DEM is the only required input in this scenario. More advanced models will include other variables, such as subsidence, tide gauge records, and datum adjustments; you should treat this analysis as a visual tool to guide planning.

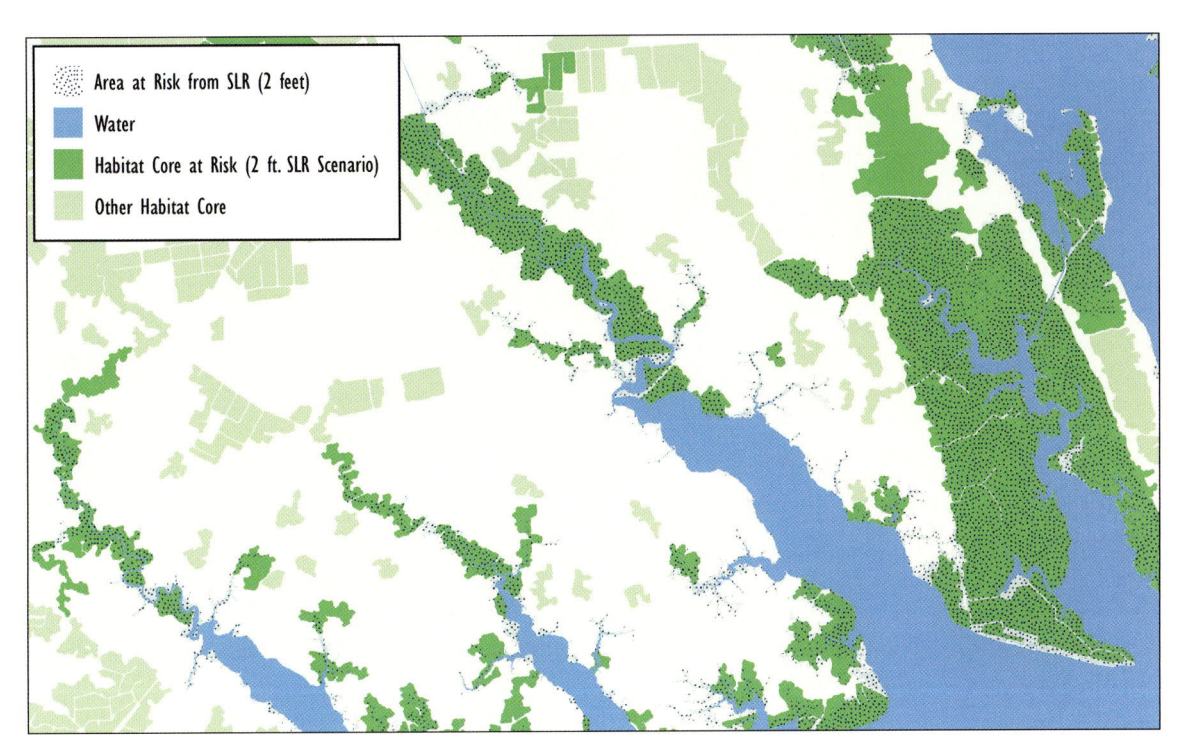

Cores at risk from SLR can be identified by simply intersecting the habitat cores layer with a sea level rise model. This map uses a 2-foot rise in sea level, and cores intersecting this zone are shown in a darker shade of green. Credit: Esri, NOAA.

Once the layer is created, a simple selection, such as "all cores that intersect the SLR layer" can be highlighted. Your next question, which will require expert consultation, would be, "Is the core disappearing or changing its nature?" In other words, is the core changing from a coastal forest habitat to a shallow marsh habitat? If it remains large enough to provide adequate habitat, it would still be a core—but a different kind.

Modeling SLR risk will also highlight other needs. For example, a greenway trail through a spartina marshland may need to be relocated father inland or converted to a boardwalk. Similarly, rising seas may take over a public use, such as a beach or parking lot located on a sand dune, requiring the acquisition of land farther inland to ensure that marshes have room to migrate and public land acreage is maintained.

The way to select opportunities presented by changing water levels is discussed further in chapter 7. Remember that SLR may not endanger wildlife and other upland species if they have ample natural land to accommodate evolving shoreline boundaries. Similarly, if SLR will not affect structures or facilities, then the impact may be considered minimal. Some structures, such as floating docks and floating bridges, are designed to adapt to changing water elevations.

Chapter 6 | Assessing risks to your assets

In-between effects of rising seas can also cause damage. For example, rising water tables can flood parkland more frequently and slowly damage upland species. High-tide flooding, sometimes referred to as "nuisance" flooding, is flooding that leads to public inconveniences, such as road closures. Nuisance flooding can damage trees through repeated inundation of saltwater and salt spray. A high-tide event (also called a king tide) can become a flood event, and storm surges can become more frequent, more severe, or both. NASA's Earth Observatory suggests that "outcomes of an increase in global temperatures include increased risk of drought and increased intensity of storms, including tropical cyclones with higher wind speeds, a wetter Asian monsoon, and, possibly, more intense mid-latitude storms." See **https://earthobservatory.nasa.gov/features/RisingCost/rising_cost5.php**.

The damages from coastal erosion and scouring can also be long lasting. In tropical ecosystems, mudslides and slope failures are more common, because the ground becomes so saturated that pore spaces become fully filled with water.

As storms become more intense, caused in part by the excess water in the atmosphere that accumulates when rising temperatures cause greater ocean evaporation, greater damages are possible. In 2017. Hurricane Harvey deluged the Texas coast with 27 trillion gallons of water, more than four times the 6.5 trillion gallons that Hurricane Katrina dumped in 2005.

Risks not easily revealed by maps

Some risks will emerge only through consultation with local, knowledgeable experts. Not all risks are mapped and not all necessary data are readily available. Users should consult local experts before, during, and after data accumulation because they may know of plans for an area that are not yet on paper. In one example, the authors working with a Virginia county were made aware of an owner who had purchased many tracts of land in one particular valley and had constructed "driveways" large enough to serve as entrances to major subdivisions across multiple plats. Although there was no pending subdivision, the owner's intentions were clear—clear enough that a user could hand digitize the area at risk to include the likely future development.

In another case, a core site was revealed to be an old dumping ground. From aerial imagery, it appeared forested, because forest had grown over the dump site. To be considered as habitat, the site may need to be cleaned up to provide a safe area for animals or people. Furthermore, although some old landfills are mapped, many landfills are known only to locals and may also need to be hand digitized.

This chapter explained the many risks that can affect a GI network, from development to fire to natural disasters. Some of these risks can be determined by reviewing GIS layers (e.g., road proximity), whereas others (e.g., development plans) require consultation with other entities, such as local governments. Risks have a hierarchy ranging from urgent to long term, and you should consider the timing and intensity of risks when you formulate a strategy. Chapter 7 is perhaps the most important chapter, because it discusses what to do and when to do it as the crux of any GI planning effort. Not all risks are equal, not all need immediate attention, whereas others can wreak havoc on the landscape.

There may also be risks that cannot be overcome, such as rising sea levels, but the knowledge of those risks should drive prioritization and conservation efforts. Mapping and evaluating GI should drive the most strategic decisions possible. By creating these maps, a project can be effective in achieving real goals, which will be discussed in chapter 8, on implementation.

Notes

1. Bentz, Barbara J., Jacques Régnière, Christopher J. Fettig, E. Matthew Hansen, Jane L. Hayes, Jeffrey A. Hicke, Rick G. Kelsey, Jose F. Negrón, and Steven J. Seybold, "Climate change and bark beetles of the western US and Canada: Direct and indirect effects." *BioScience* 60, no. 8 (2010): 602–613. http://www.treesearch.fs.fed.us/pubs/36133. Website accessed January 2018.

2. Julie Cart, "In Lake Tahoe development struggle, California blinks, Nevada wins." *Los Angeles Times*, September 14, 2013. http://articles.latimes.com/2013/sep/15/local/la-me-tahoe-development-20130916. Website accessed December 2017.

3. Lesk, Corey, Ethan Coffel, Anthony W. D'Amato, Kevin Dodds, and Radley Horton, "Threats to North American forests from southern pine beetle with warming winters." *Nature Climate Change* 7, no. 10 (2017): 713.

4. "Nine species that are feeling the impacts of climate change." Department of the Interior Blog. https://www.doi.gov/blog/9-animals-are-feeling-impacts-climate-change. Website accessed December 2017.

5. Epps, Clinton W., Dale R. McCullough, John D. Wehausen, Vernon C. Bleich, and Jennifer Rechel, "Effects of climate change on population persistence of desert-dwelling mountain sheep in California." *Conservation Biology* 18, no. 1 (2004): 102–113.

6. Charmantier, A., and P. Gienapp, "Climate change and timing of avian breeding and migration: Evolutionary versus plastic changes." *Evolutionary Applications* 7 (2014): 15–28. doi:10.1111/eva.12126.

7. Nussey, D. H., E. Postma, P. Gienapp, and M. E. Visser, "Selection on heritable phenotypic plasticity in a wild bird population." *Science* 310 (2005): 304–306.

8. Lewis, Renne, "Climate change played role in Everest avalanche, scientists say." *Al Jazeera*, April 26, 2014. http://america.aljazeera.com/articles/2014/4/26/everest-climate-change.html. Website accessed December 2017.

9. Mears and Wilber, "Identifying avalanche terrain." http://mearsandwilbur.com/terrain.html. Website accessed November 2016.

10. Rudolf-Miklau, Florian, Siegfried Sauermoser, and Arthur Mears, eds., *The Technical Avalanche Protection Handbook* (imprint unknown, 2014).

11. Pierson, T. C., R. C. Evarts, and J. A. Bard, *Landslides in the Western Columbia Gorge, Skamania County, Washington*. US Geological Survey Scientific Investigations Map 3358, scale 1:12,000, pamphlet (2016). http://dx.doi.org/10.3133/sim3358. Website accessed December 2017.

Chapter 7

Determining opportunities

Green infrastructure (GI) planning aims to systematically recognize highly valued assets and protect them. Earlier chapters discussed what to prioritize and why one core could be more important to focus on than another as an integral part of a GI network or conservation plan. However, it is also possible to restore assets by replanting impaired landscapes and by removing or abating impacts, such as mine spoils and fences that impede wildlife movement.

Recall that mapping GI assets is not the final stage of your work. Understanding what may change is critical to informing actions. Earth can survive without people, but people cannot survive well without a healthy planet. This chapter discusses what to do next—how to determine whether action is necessary and what steps to take immediately to protect a landscape when risk modeling shows that such action is required soon to avoid or overcome the risk. Indeed, perhaps the most interesting aspect of mapping is using maps to surface opportunities. In other words, you can address ideas revealed by GIS analysis and consultation with experts and stakeholders.

At first, idea generation should be as expansive as possible. Then you can prioritize which ideas are most important and eliminate the rest.

Prioritizing opportunities

The first step in surfacing opportunities is to identify and assess risks, as previously explained. The GIS lead (and the advisory group or local committee) can then relate those risks to the goals of the project. However, because GI planning is an iterative process similar to adaptive management, risk assessment may have suggested new goals. For example, a core or corridor that was not previously a priority may be found to be at risk. The committee may not have identified protecting a drinking-water supply area as a goal until the risk assessment showed that growth pressures were highest in the area closest to the recharge area.

Overlaying the risk analysis on to the base map of cores is a critical step that further informs a GI plan, especially in assessing what (if anything) can or should be done. This step will show where cores of high value (or that occupy a key placement in the network) are at risk. The risk assessment determined why it is at risk and how imminent and severe the risk is. The question now becomes, "Is that risk unavoidable, or can the project mitigate or remove it altogether?" Rather than seeing risks simply as problems, consider that they are also opportunities to integrate development into a wider GI plan.

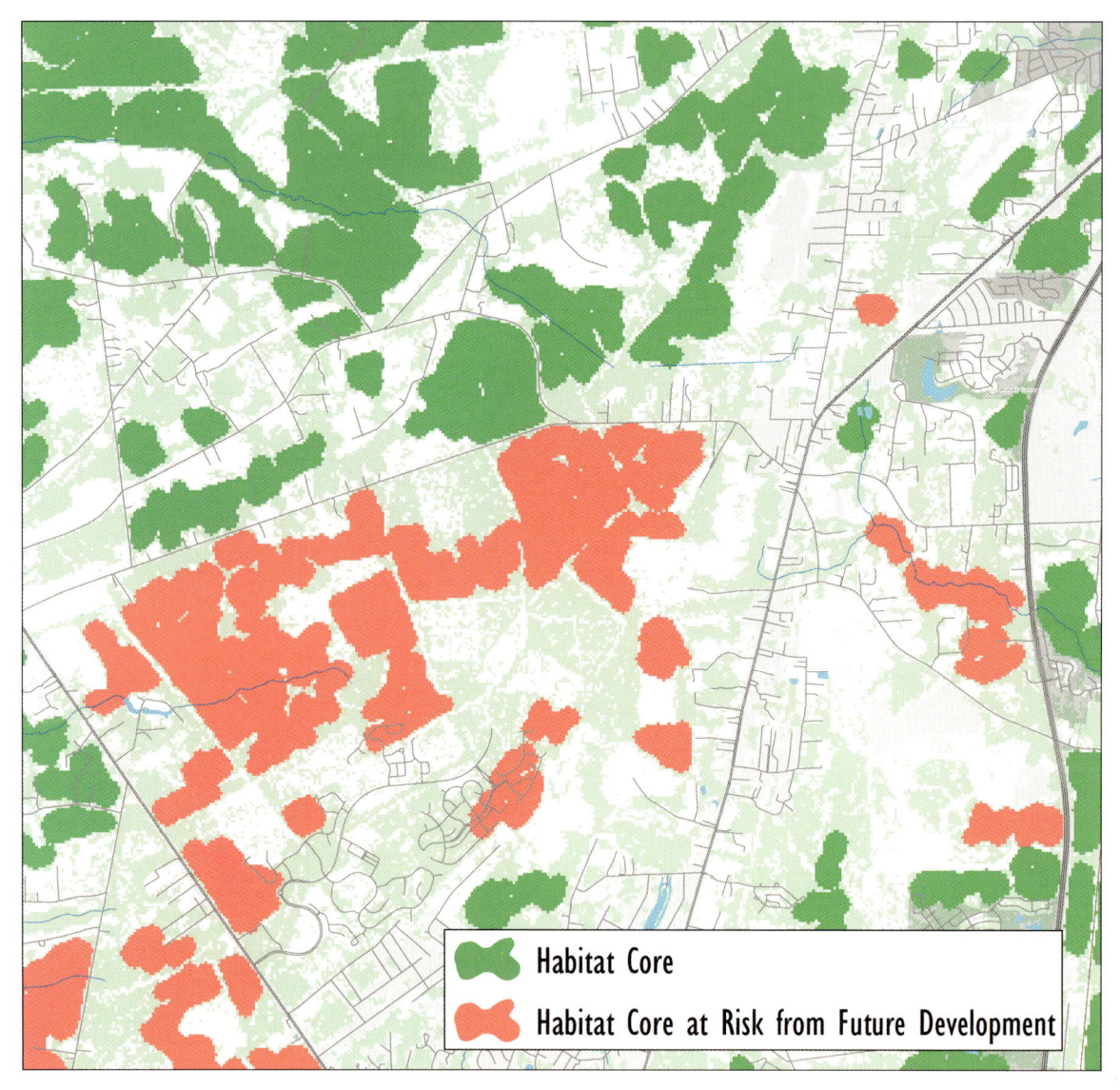

Flag cores that may be at risk and consider whether they present opportunities for action. Credit: Green Infrastructure Center Inc.

First, you must prioritize the relevant risks. Then you rank them in terms of their significance to the goals of the GI plan. For example, if you identify 30 risks to the county's drinking-water supply, which risks are the highest priority? Focus on them and dismiss other risks that do not affect the GI goals.

Sometimes, risks cannot be changed. For example, a development project that has already received all its approvals from the governing body is not likely to be stopped. Similarly, a new road, powerline, or gas pipeline that has already been approved and funded is unlikely to be stopped altogether. It might be rerouted slightly to avoid an ecological impact such as a wetland or a cultural artifact such as a historic burial ground, but once the permits have been issued, these types of projects are almost impossible to rescind. They may, however, need to mitigate impacts, which could provide funds for conserving key cores elsewhere.

Opportunities at the level of independent parcels

Another consideration related to risk abatement is property ownership. If a core under private ownership is likely to be developed, how amenable is the property owner to considering ways to protect and conserve the core—at least in part? If not, will the owner consider selling the land or placing it under a conservation easement? Will the developer consider relocating a project to another part of the property or to the edge of a core instead of its center? Is it too late to rezone the land to stop its development?

At the parcel scale, consider the following options:

1. Purchase the land and put it under easement to protect it in perpetuity.

2. If development threatens the drinking-water supply, a locality can purchase it to protect that supply.

3. Work with the landowner to relocate or redesign the development to maximize retention and protection of the core.

4. Decline to service the development with sewer lines or new roads.

5. The extension service, a local forestry agency, or a conservation group could work with the landowner to ensure wise management of the parcel for habitat conservation or to prevent stormwater runoff beyond the minimum requirements.

6. If the risk threatens clean water supplies, persuade nearby landowners to redesign a development so that most of the land—especially land nearest the reservoir or stream—is left undeveloped by allocating it to dedicated, vegetated open space. Taking this step would reduce the development footprint and allow strategies to reduce runoff throughout the development.

7. Change the zoning of the entire core, or, if a development is already approved, of the parcels nearest the water body by adding a zoning overlay to impose stricter land management requirements for protection. (Courts in the US have not found this form of overlay zoning to be a taking, because it does not remove all economic use of a parcel and protects a public resource.)

Zooming out to the macro scale reveals other opportunities, which we explore next.

Larger-scale opportunities

When changing scales, cores may become more significant as part of a larger regional network. They may fall within an international migratory network or be part of a major river floodplain, such as the Mississippi River floodplain. They might form part of an interstate trail, such as the Continental Divide National Scenic Trail or the Natchez Trace National Scenic Trail. Examples of migrations that could fall across a large area (discussed earlier in this book) include elk migrations, hawk migration routes along ridgetops, and bighorn sheep pathways. Animals commonly use some areas during winter to move down to lower elevations to forage. The locations of these areas may be determined by consulting with local or regional conservation groups, state wildlife biologists, and others.

In the US, state wildlife or game agencies create wildlife action plans. Only recently have these plans been spatially represented (as opposed to written lists of species by county). If the data are not readily available in the plan, the agencies can be contacted to inquire whether there are known migratory routes or other key habitats that could be shared with them and presented in a spatial format. However, sometimes the data are at such a coarse level that they are almost useless. As noted earlier, bear habitats can cover an entire county, and some bird reserves, such as for the whooping crane, cover swaths hundreds of miles wide. Depending on the area of interest, these maps (which are sometimes not accurate) may be less than helpful because they encompass an entire locality or region and do not show distinct priorities—the entire county is a priority.

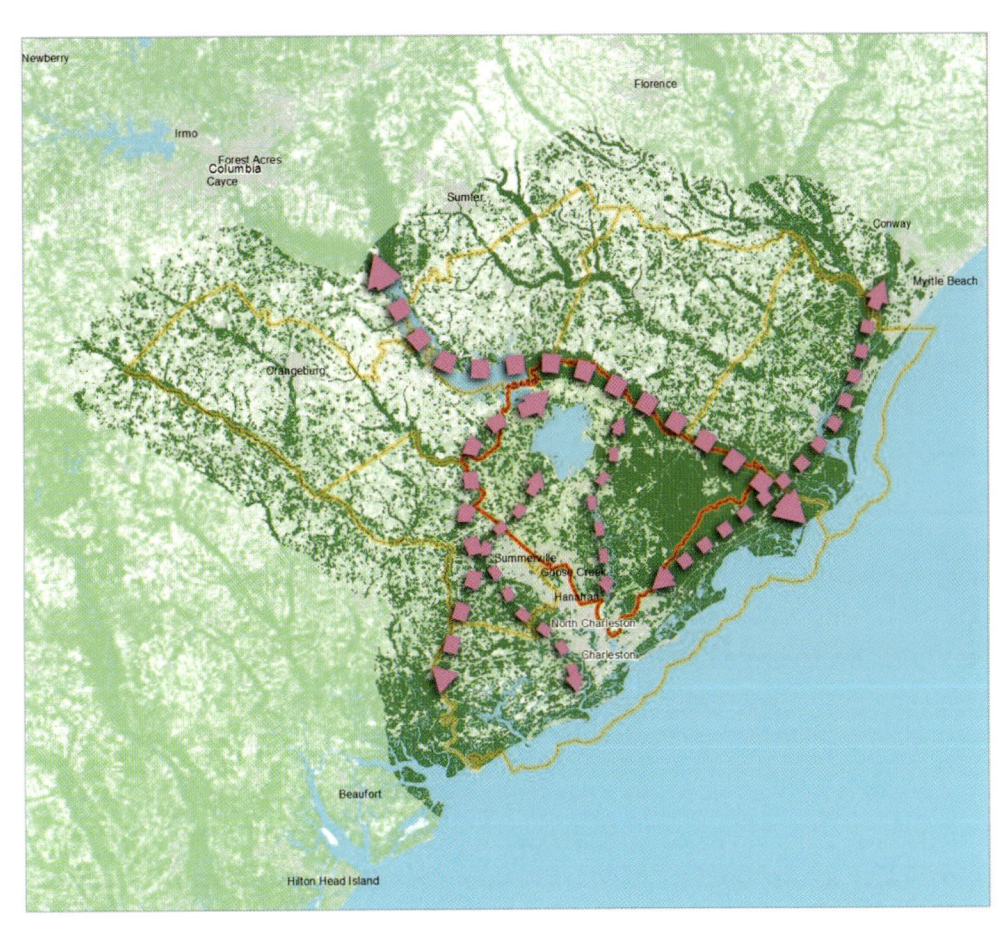

Consider key regional connections and the local cores that connect them as critical when seen at a larger scale. Credit: Green Infrastructure Center Inc.

Assessing rarity

One rationale for conservation is to protect an area because it is a rare example of a remnant landscape. For example, if a project goal is to protect the area's biodiversity, then selecting cores that are known to support a rare plant or amphibian, or have remnant plant communities from before the last Ice Age, could be another way to recognize areas for protection.

However, conservationists should not limit a GI network only to areas that support rare species. It is more strategic to protect high-quality habitats that support many species. As one state wildlife official said at a community meeting, "Don't just try to conserve rare species. We should also keep common species common. It takes tremendous effort to bring an animal off a state or national endangered species list." Ensuring there is a robust network of habitat cores rich in diversity is the best way to ensure that species do not become rare in the first place.

A plant, mammal, fish, or amphibian may not be globally, nationally, or even regionally rare, but it may be locally rare and thus warrant conservation action to protect its habitat. When it comes to rarity, a core could be significant to a community because it represents a type of habitat of which little may remain in the future. Such a core may be of average quality in terms of richness, size, and diversity but provide a key stepping stone to link two cores or be a unique habitat type that is locally rare.

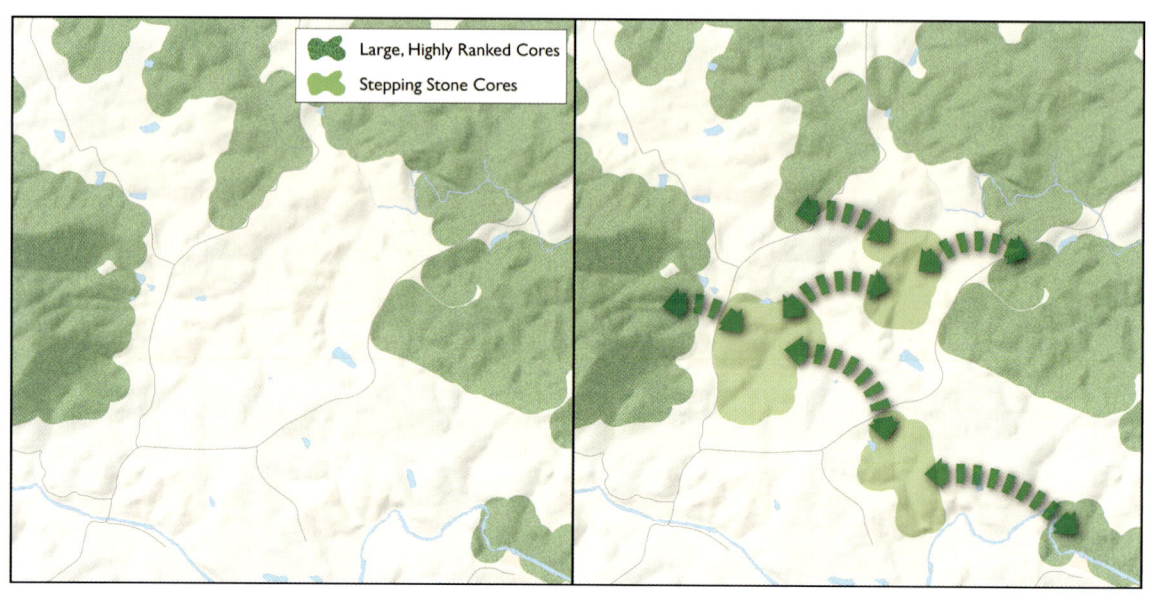

First select the highly ranked cores. Second, add lower-ranked stepping stone cores to provide passage for wildlife and people. Credit: Green Infrastructure Center Inc.

Assessing restoration potential

Restoring a natural landscape means restoring its richness, diversity, and connectedness. Although taking a landscape-wide approach is vital, often the restoration of an entire landscape requires step-by-step measures, restoring first one core and then another, adding a corridor or stepping stone here or there—in other words, taking a piecemeal approach within the context of a general approach as funds and staff resources allow. This process also allows the landscape to accommodate parcels of development within it.

A GI plan rarely claims, "We must preserve everything at all costs!" Rather, the plan identifies key habitats and connective elements for preservation or restoration. Because many GI projects encounter fragmented landscapes, the GI planners are challenged to identify key habitats that need restoration. A successful GI strategy often includes protecting existing natural assets, improving their quality, and adding to their extent. When reviewing a map of existing natural assets, the analyst may find disconnected or degraded areas. If two habitat cores lack a connection, a new corridor could be planted or a parcel purchased to provide a stepping stone between them. Similarly, a forest or wetland core could be expanded by planting more trees or removing invasive vegetation. The plan can encourage landowners to protect the core by saving them from paying higher taxes for a use zoned for development or by giving them other benefits.

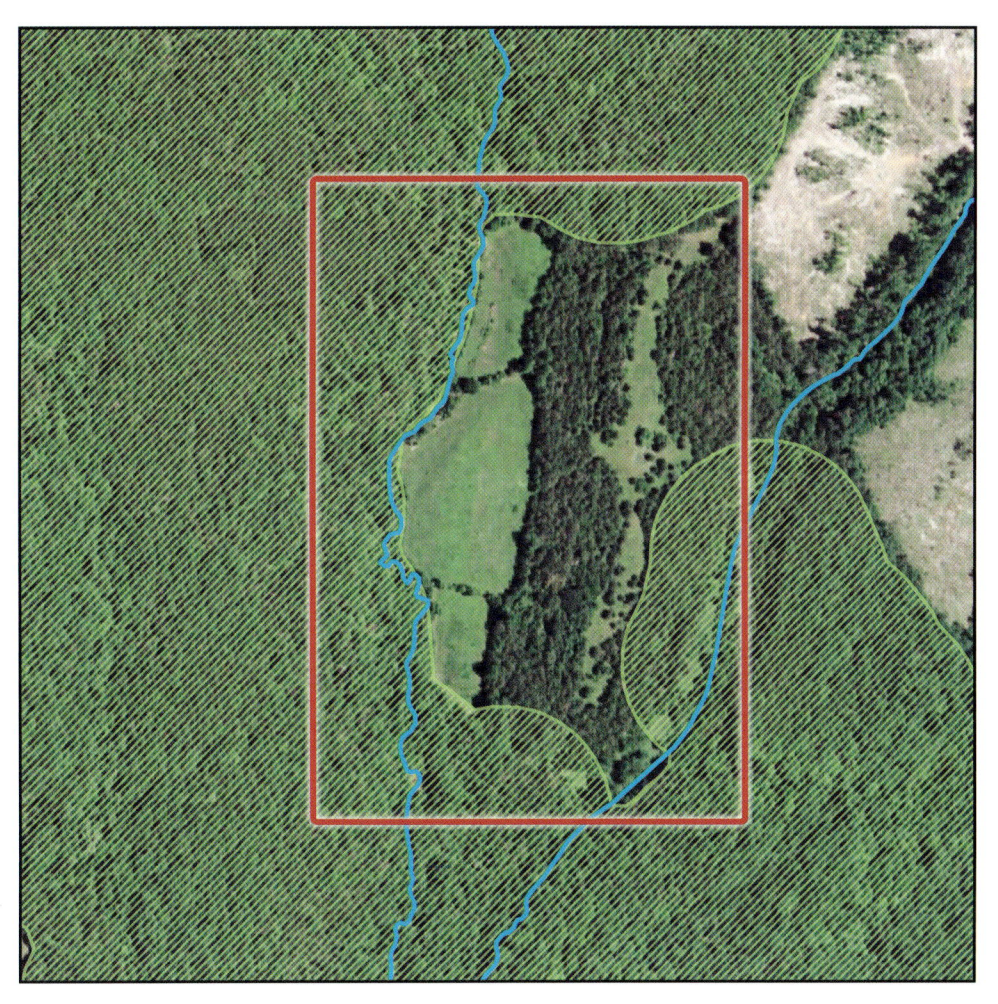

Habitat cores, shown with crosshatching, surround a fragmented landscape. Those nonforested areas could be filled both to enlarge the core's habitat and to better protect the stream. Credit: Green Infrastructure Center Inc.

The Cherokee National Forest in Tennessee offers an example of habitat restoration to meet a specific conservation goal. The forest was modeled to determine what clusters of forest types were present and to examine how past alterations had caused deviations from what might naturally occur. For instance, a plantation forest had been annexed into the national forest, resulting in an anomaly from a naturally occurring forest. Rather than a diversity of naturally occurring forest types, a single forest type of closed canopy, with a low diversity of trees, dominated some areas.

One of the stakeholder groups involved in the conservation planning project for the forest was the Ruffed Grouse Society. The ruffed grouse is a chicken-sized bird that blends in with its forested habitat and is a popular game species. It has the distinctive behavior of "drumming" to attract a mate. The society wanted to add more forest openings for access to forage, sunlight, and herbaceous plants. Members asked for the forest plan to create some small clearings throughout the forest—more than might normally occur to compensate for grouse habitat destroyed elsewhere. The group seized upon this opportunity to mitigate past damage from clearing forests elsewhere.

Seeing transportation plans as opportunities

Transportation plans are usually regarded as risks to natural assets. They tend to disregard the needs of wildlife and fragment cores because of a different set of priorities than the preservation of natural habitats. Road designers tend to see mountains, swamps, and rivers as obstacles to overcome, not as opportunities to facilitate wildlife habitats. However, GI planners can see road projects as opportunities to integrate a community's goals into its transportation plans. Decision makers can use the GI cores map to inform changes to currently planned roads and planning for roads not yet designed. Mitigation funds required for many road projects can be used to save cores elsewhere.

To find transportation opportunities, analysts should ask the following questions:*

- Where are future roads planned? Will planned roads bisect natural features that have been identified as high priority?
- Are the roads needed? For example, are transportation-demand models based on current population projections or on a suburban model with excessive numbers of daily trips that does not best represent the community?
- Could other less impactful routes be considered?
- Can any new development stimulated by a new road be directed away from important natural features?
- Would alternative transportation models solve some of the demand to move people?
- If road projects need to purchase land to mitigate impacts, such as wetlands or open space, can the natural asset map prioritize which lands to acquire?
- Can new roads incorporate new approaches to green highway design that are less impactful to wildlife or provide medians for trees?

* If possible, obtain transportation data in GIS and overlay them onto the cores map.

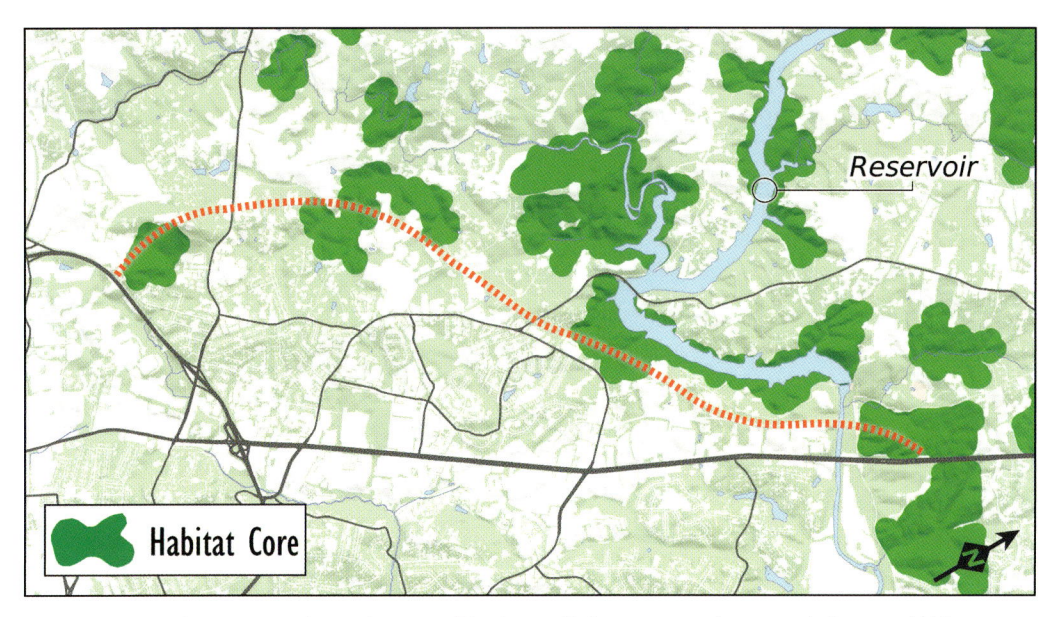

A large highway bypass was planned next to Charlottesville's water supply reservoir that would bisect habitat cores that protected it. Credit: Green Infrastructure Center Inc.

The Piedmont region of central Virginia offers an example of how a new road—one that was not completely planned and approved—was modified to protect a water supply. A large highway bypass was planned next to Charlottesville's water supply reservoir that would bisect habitat cores that protected the reservoir. Although the road had been designed, approved, and funded, a change in local governance resulted in the land being returned to its established uses. The new board of supervisors deemed the road unnecessary and overly destructive to the water supply, schools, burial grounds, and other valued natural and cultural assets. This decision saved the reservoir and surrounding forest cores.

The US Department of Transportation published a useful book, *Eco-logical,* to provide guidance for how to design roads that minimize disturbances to wildlife.[1]

Impaired landscapes, such as landfills, brownfields, and dump sites

Opportunities may exist to expand the network of landscapes of high value by restoring areas that have been degraded through a variety of human activities. As land can be restored by removing contaminants, replanting native vegetation, removing invasive species, and other approaches, analysts should consider these sites as opportunities for restoration. Often, such sites are in the middle of urban areas or along riverbanks, so their restoration will have high value for human leisure and recreation activities and will create new urban cores and restore old riparian corridors.

Impairments may be difficult to determine using only the GIS cores model, and users should remember that the Esri model does not access local data. As part of risk assessment, known brownfield sites may have been added to the map, along with old dump sites and other contaminated areas. Some toxic dump sites become overgrown as woodlands and wetlands reclaim old factory, extraction, and landfill sites. Although these sites are inherently impaired, they may easily show up on satellite imagery as core habitat of high value. This is the reason it's important to add local data for leaking underground storage tanks, old mine tailings, and quarry sites that may show how an area that looks pristine suffers from a past chemical spill, piles of rusting machinery, or a toxic dump.

If these areas do exist, planners should consider whether these sites should be under active cleanup as part of a wider habitat restoration program. Some features, such as old mining slag heaps in Kentucky and West Virginia, have revegetated naturally, and it would be too costly to remove them. In contrast, a site that is leaching contaminants should definitely be cleaned up, especially if there is the possibility of impacts to humans or animals.

Many localities have repurposed impaired sites, such as landfills, as parks. Often unsuitable for buildings because of uneven settling, localities can reclaim landfills by adding clean fill dirt, channeling leachate to treatment areas, and trapping off-gassing methane and other air contaminants. A landfill could be stabilized, replanted, and reclaimed as a habitat core or possibly a key stepping stone in a wider network. Gas emissions can be captured to generate power.

Strategic location of cores in meeting goals

GIS can help planners derive many statistics to inform conservation actions and evaluate how well a goal has been achieved. However, some spatial statistics could be misleading—or not tell the whole story—when considering the strategic arrangement of a green network. Although a GIS analyst could choose a geography of interest, say, a county, and run statistical analysis on it, such as total acreage of cores in the county, the result may not tell the most important story.

For example, what if all the cores are in the western half of a county where elevations are higher and steep slopes make that area less suitable for most species? In this case, saying that half the county is covered by large habitat cores can be misleading. The lower elevations may have more species diversity, and certainly they will be different species and even dramatically different ecosystems. The eastern half of the county might be piedmont or tidewater, meaning that the cores in that area are more at risk while having the greatest value for species' conservation. The county may be "cores rich" overall, therefore, but many of its species still need protection. The greatest opportunity to meet conservation goals may be in the area that has the greatest risk and the most fragmented cores.

Following are some useful questions for analysts to ask to surface new opportunities. In these cases, it's important to reconsider the project's goals to recall what results they are trying to achieve. These questions are examples only. Analysts should frame their own questions to reveal new opportunities. The questions are framed with those goals developed earlier in mind.

Wildlife goals

- Will some cores support a species of interest to the community or group responsible for the GI plan?
- Which areas are known to support multiple species?
- Which areas are at greatest risk (from earlier analysis)?
- In what areas will human and wildlife needs conflict? Would additional buffers be useful to protect a breeding area or allow for migration without conflicts with roads?
- Will some areas lend themselves to wildlife watching, viewing platforms, and nature walks (to support tourism and environmental education)?
- Will some areas lend themselves to species reintroduction?

Overlay known wildlife habitat needs for species of interest. For example, one species might be limited to land above a certain elevation, such as the American pika (*Ochotona princeps*), whereas an animal such as the Sonoran Desert tortoise (*Gopherus morafkai*) occupies only lower elevations. The richest cores tend to cross a wide range of elevations to capture both species. For example, in the American Southwest, individual mountain ranges—often referred to as "sky islands"—act as islands of diversity and life within a wider desert context.

If the group has concerns or would like to achieve management goals for a particular species, it can map and overlay these needs (elevation constraints, land cover, forage, abundance of prey, water sources, symbiotic species, and so on) for each animal. The result is a composite map showing where multiple species needs intersect most advantageously. For an overview of such approaches, see Beier, Spencer, Baldwin, and McRae.[2]

The American wolf, bison, and elk are three examples of large mammals that disappeared from many parts of the US, mostly from hunting, and have now been reintroduced into areas where they were once abundant. Identifying areas to support reintroduction of species could be useful, especially as they tend to attract tourists.

As wildlife habitat may need protection, consider where conservation easements are located so that they can support more permanent open space for wildlife habitat and migration. Some cores or corridors may need restoration to achieve this goal. For example, cross-county connectors (key corridors) were identified for Accomack County, Virginia, in two ways: cores that are currently good connectors and corridors that could lend themselves to restoration to become good corridors.

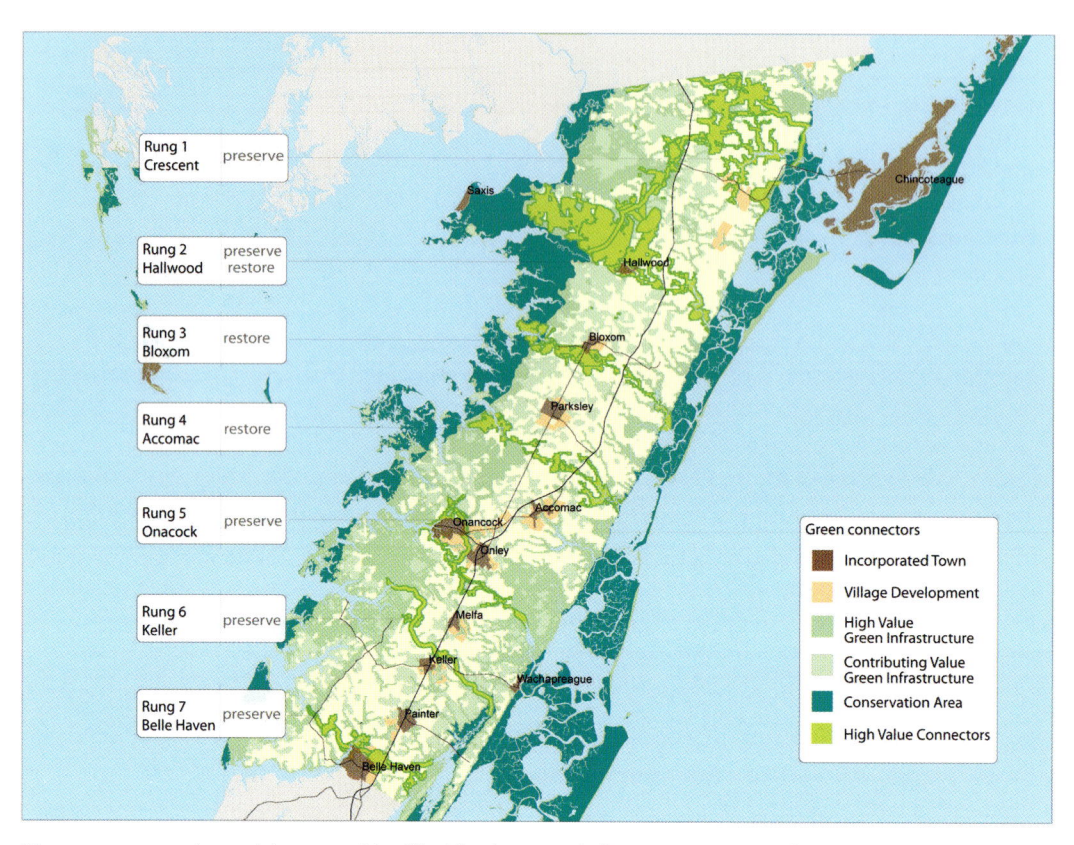

These cross-county corridors were identified for Accomack County as conservation or restoration potential. Credit: Green Infrastructure Center Inc.

Once these cores or corridors have been identified, the next step is to seek a conservation easement, create protective overlay zones, and obtain grants or landowner cooperation to replant the area needing restoration. Subdivision planners can cluster homes away from the corridors and generally discourage development that would block wildlife access to and movement through the corridor.

Water quality goals

- What percentage of streams intersect cores?
- Is most of a river's course within a series of cores? If not, should some stretches be restored to complete the riparian corridor? If floodplain data are available, select all cores that are within or intersect at least 50 percent of the floodplain.

- Do cores protect the headwaters? Use GIS to intersect all first-order streams with habitat cores to select for those that support a headwater core.

- What percentage of the river's floodplain is covered by cores (or forest cover, even if somewhat fragmented)? Clip and intersect the floodplain with cores to obtain the percentage of floodplain covered by habitat cores.

- Which reaches of streams are impaired, either because their habitat is damaged or by pollutants and stormwater runoff? Can cores be planted or expanded to enclose more of the stream, or should cores closest to streams be prioritized for protection to permanently buffer them?

- Are there opportunities to protect cores with a large percentage of erodible soils (found with soil survey data) or steep slopes?

If habitat cores occur only within the lower reaches of a river, while the upper reaches are unprotected, the lower stretches likely will suffer water quality impacts, regardless of how well they are protected by nearby cores. Often, headwater streams are more sensitive to impacts from land clearing or runoff because their drainage is smaller, their slopes steeper, and their soils thinner. Protecting them is vital for downstream water quality.

Furthermore, cores at a higher elevation may support rare species not found at lower elevations. Overall biodiversity may be lower in headwater streams, especially in terms of fish and plant life, because the upper reaches of a river have fewer nutrients. As streams move downslope, more biological matter enters them, supporting a richer assemblage of organisms and a more diverse food web. However, impacts such as stormwater outlets, golf courses, and factories tend to reduce species diversity in the lower reaches as streams pass developed areas.

You may want to consult water quality data obtained in step 2 (reviewing the data), which may reveal sections of water courses that support rare species and the actual conditions of the stream, such as its dissolved oxygen content. Streams change over their course, and different influences could come into play. For example, a reach of river might flow through a unique geological formation, giving way to an altered water chemistry, which in turn supports a new amphibian or provides key habitat to a more successful fishery. Aquatic ecologists can offer advice if they are on the review team for the mapping effort.

Remember that in reviewing land cover and relating it to goals for water quality, you may encounter pipelines under forested areas. These pipelines may deliver pollutants directly to a stream from urban areas or other pollution sources such as mines or landfills. Thus, a river could suffer from piped discharges, despite forest cover along its banks. Similarly, a river originating outside the study area may carry pollutants from upstream that impair water quality, so the river may be impaired even if the habitat core system is in excellent shape.

Water supply goals

- How well protected is the waterway (or watershed) above a water intake pipe (e.g., a pipe that draws water from a river to an offsite reservoir and will be used for drinking water)?

- For a reservoir, how well does the land cover protect the immediate drainage (within 1,000 feet of the waterside) and the watershed generally? What percentage of that area is in cores?

- If much of the area relies on groundwater (and on rivers supplied by groundwater), could the water recharge areas be better protected? Does native vegetation protect the aquifer recharge area to facilitate rainfall absorption?

The danger of developing water recharge areas

A key challenge when planning for water supply is that areas where significant recharge occurs are often not well mapped. For example, community members are often surprised to learn that permits for new developments that are reliant on public wells have not mapped the aquifer and do not know how much volume of withdrawal can be supported (or when it will be exceeded). Similarly, areas that are key for recharge may be unknowingly paved over. Furthermore, recharge areas may not be where the habitat cores are, because recharge pathways are governed by topography and underlying geology, not by land cover.

In Accomack County, for example, the recharge area is along the spine of the Eastern Shore of Virginia, generally mirroring US Route 13, the main road that serves the southern Delmarva Peninsula. The road follows that central spine, which means that the drainage of the peninsula is into the Chesapeake Bay on the west and the Atlantic Ocean on the east. With US Route 13 almost mirroring the recharge area, most of the development pressure is along the area that should be kept as open and pollution free as possible.

The unique geology of the Eastern Shore means that this narrow spine is the only access though which water can recharge the deeper drinking-water supply aquifer. The situation presents both a challenge and an opportunity. The opportunity is to educate planners about the water recharge area and encourage efforts to reduce impacts and development on top of the recharge zone.

This example provides an interesting reminder for GIS analysts. They are not water quality or supply experts and will rely on advice from such experts. As explained throughout this book, GI planning requires participation from multiple disciplines and iterative consultation with other groups. In Accomack County, the localities formed a groundwater committee to address their ongoing issue. The key was to obtain the data. The GIS project team mapped the information to highlight its importance. Based on that data, the information is now published so that residents can understand the situation. New zoning restrictions or additional stormwater quality and quantity controls can be implemented and other measures taken to protect the recharge spine. Only a map of these relationships could reveal these opportunities.

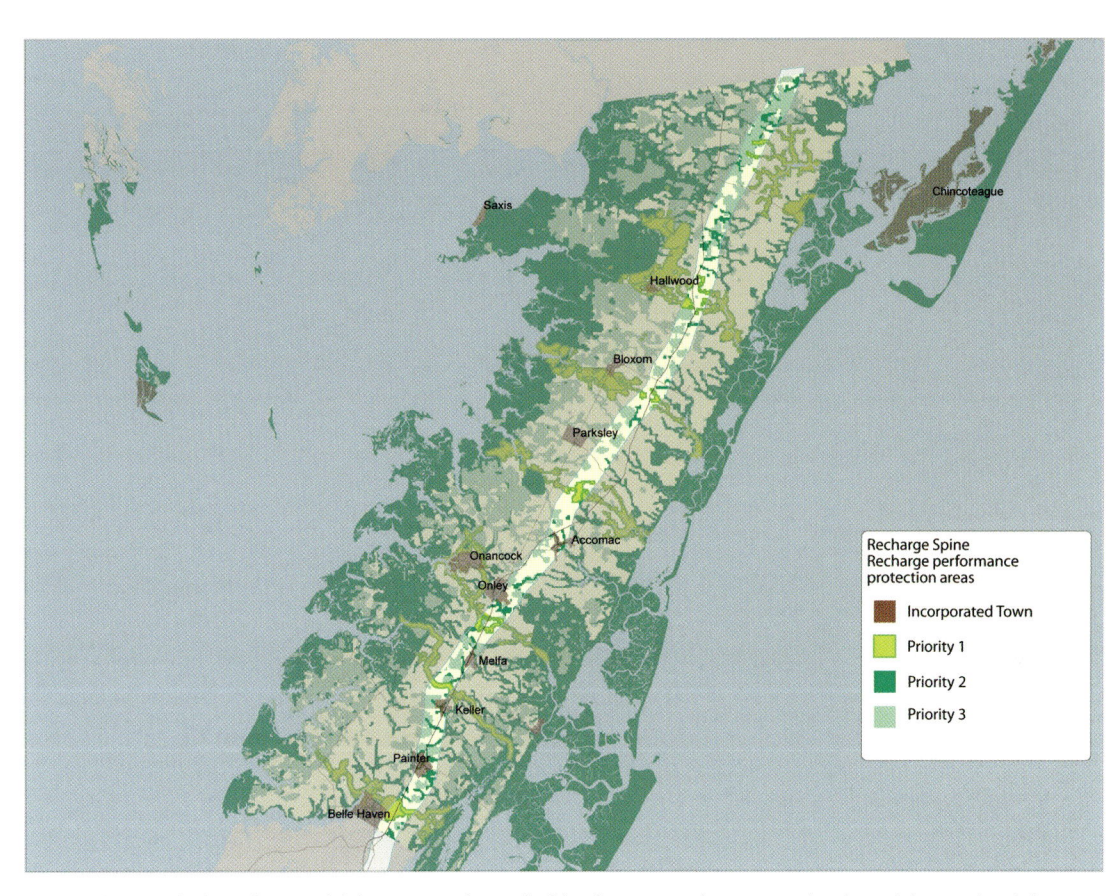

This recharge window shows which areas are key priorities for preventing contamination of the peninsula's water supply. Credit: Green Infrastructure Center Inc.

Recreation goals

- Which cores support current recreation, such as hiking, mountain biking, birding, boating, fishing, and swimming? Consider prioritizing those cores for conservation if that is a focus of the project.

- Which cores could support recreation? Which cores could provide river access if easement or ownership were obtained? Which core could be a prime spot for watching hawk migrations or proximate to a town that lacks open space?

- Which areas could support recreation if a core was restored or expanded? Could access be created by adding a trailhead or parking area? Could a greenway trail be threaded through the area if trees were planted and wetland was restored?

In GIS, overlay the recreation-themed map to show where cores support current recreation and prioritize those. Consider the flipside of such analysis. If access to nature-based recreation is important, select all cores on the map that support recreation and that lie within a certain distance of a residential area: if walkable, chose areas within a mile; if drivable, chose areas within 10 miles or so.

Next, identify areas that lack such recreational access. This identification may also suggest a strategy. For example, are there cores in the north of the county that were not ranked very highly but that would be ranked higher if they provided access to future recreation? If you want to allow access to nature near where people live, then you want to prioritize cores closer to urban areas.

Overlay recreation and cultural interests to show how habitat cores support community interests. Credit: Green Infrastructure Center Inc. map made for New Kent County, Virginia, for its comprehensive plan. Habitat cores data modified from the Virginia Natural Landscape Assessment (VaNLA).

Scenic goals

- Which cores are key parts of a scenic view? Conversely, where are the best views? If a goal is to protect viewsheds and scenic views, develop a viewshed map as one of the overlays. Views can be from below (looking up at mountaintops or ridges) or from above (looking down across a landscape).

- Which important viewsheds could adopt standards to make development areas less obtrusive? What parts of a viewable area within an iconic viewshed are threatened by development? Could those areas be encouraged to adopt camouflaged roof colors, or could buildings be arranged to take advantage of the landscape or orientated to blend in better?

- Can scenic roadways be buffered? For flatter landscapes, clip an area along the scenic road (e.g., 100 feet from a roadway edge). What percentage of that area makes up habitat cores? For areas with mountains visible from the road, what areas contain habitat cores?

> **Protecting scenic areas**
>
> As you identify scenic areas (within key viewsheds or view areas, such as overlooks), consider what actions you will need to take to protect them. For example, are mandatory restrictions, such as limiting the height of cell towers, possible? Some localities limit the height of cell towers to the height of the tree line, even though this may require more towers overall to get the same coverage.
>
> A locality could create an overlay district for a key viewshed that restricts the height and the type of things that are "viewable." For example, a barn silo may be considered okay in an agricultural district, whereas a cell tower would not. A community committee or advisory group can determine what items to limit or screen and what height limits might be appropriate.
>
> These restrictions do not have to be mandatory. In Albemarle County, Virginia, the locality did not want to impose viewshed protection laws, but developers were strongly encouraged to work with the key stakeholder, Monticello, home of President Thomas Jefferson, to protect the viewshed from the home. Commercial buildings camouflaged their roofs, and companies, such as solar farms, carried out studies to ensure that their solar panel arrays were visible only from a few points. The county also imposed strict screening requirements for cell towers, preventing them from being taller than the local treetops.
>
> In addition, you can identify areas within the viewshed that are under easement. A local land trust could be encouraged to seek easements for areas that are within a viewshed to ensure land is protected permanently.

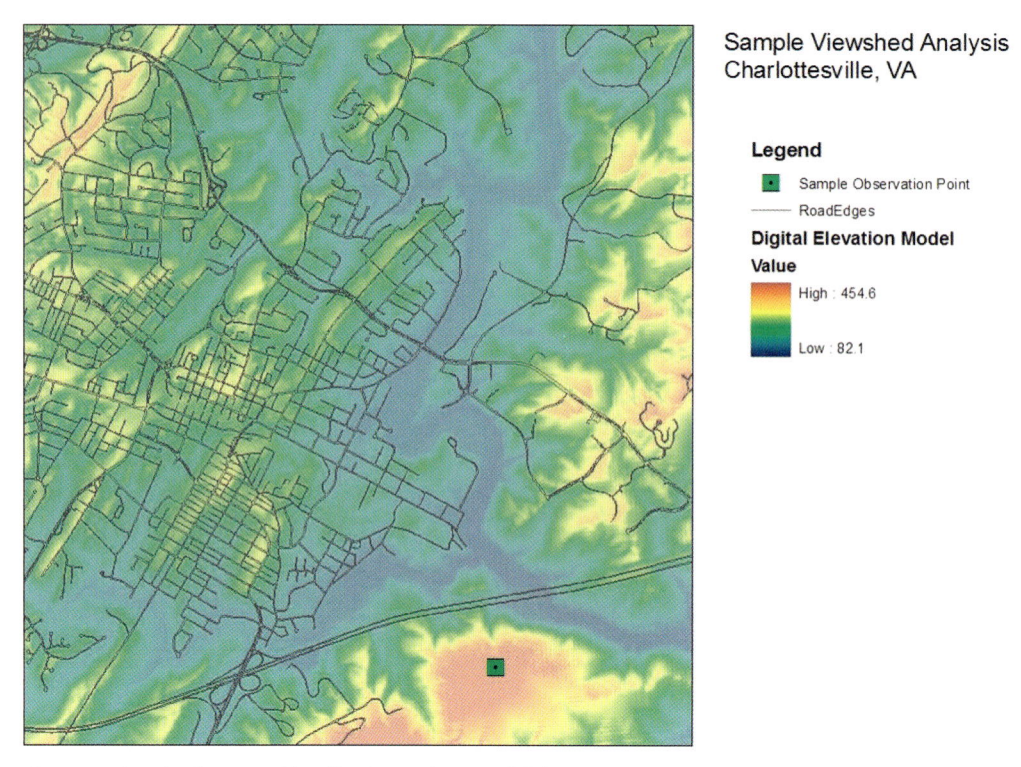

Create a viewshed map to identify places that are visible from certain vantage points. Credit: Green Infrastructure Center Inc.

Determining and proposing alternative development scenarios

In the risk map, parcels subject to development are already flagged. One challenge to the accuracy of risk maps is the "ghost" subdivision, parcels that are not yet developed, even though they have a legal right to do so. These parcels may be known, but some development rights may not be obvious. As noted earlier, parcel layers from the locality can show which areas have already been subdivided and are at risk, even if there are no structures currently on them.

However, some older subdivisions may no longer be developable because they don't meet today's standards for lot sizes; lack primary road access, water service, or adequate aquifers; or were laid out on overly steep slopes where development is no longer allowed. They may also face utility constraints such as lack of access to public water, sewer, or power. Before tagging parcels as "at risk," check them for conformity to existing development rules or access requirements (e.g., requirements to front on a public road).

A parcel may not get developed as proposed, even if zoning, roads, and utilities are already in place and the parcel is subdivided. In the next example maps, three towns in rural Accomack County, Virginia, on the Delmarva Peninsula are slated to eventually fill the green spaces between them to develop in a more compact pattern. This plan is seen as "smart growth" because it will utilize existing road networks, facilities, and utilities.

The question for a conservation planner is, "Can existing habitat corridors be maintained as towns develop?" The answer is "yes," if zoning tools such as clustering—constructing buildings closer together and on smaller lots—are applied. In the next example, a key core contains a stream and is within the area that may develop. This habitat corridor had two parcels that could be built on (A and B). If fully developed, these parcels would have narrowed the corridor and cores significantly. By building on smaller lots, the developments can accommodate more houses and conserve most of the corridor!

Using the county's own development rules for clustering in Village Residential Districts, together, parcel A and parcel B would yield 56 units, but by using smaller lot sizes, the parcels could yield 84 units while still maintaining the green corridor.

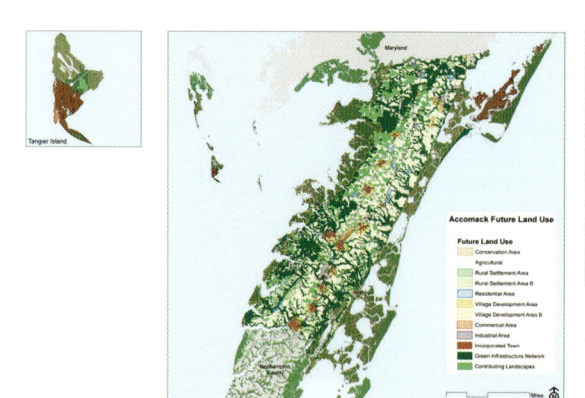

This future development map shows where Accomack County proposed to set future development.

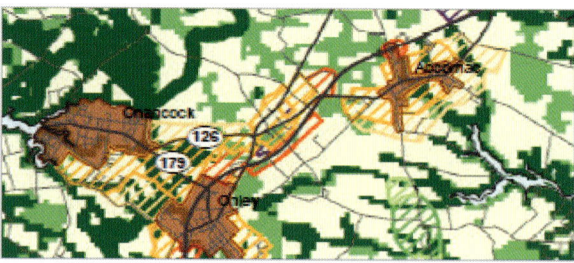

This zoom-in shows in orange crosshatch areas where infill would occur in the future to join small towns together. This plan avoids sprawl by building close to development areas. However, this could break up existing key habitat corridors.

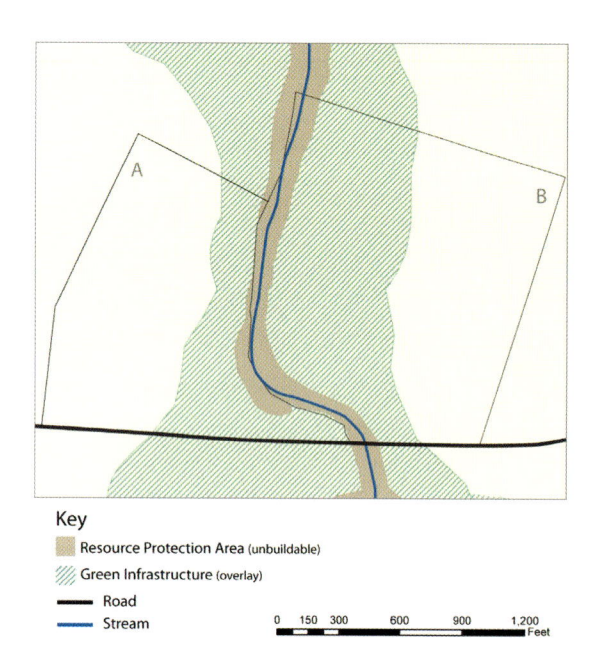

This shows an actual habitat corridor protecting a stream. Only 100 feet on each side of the stream was restricted from development.

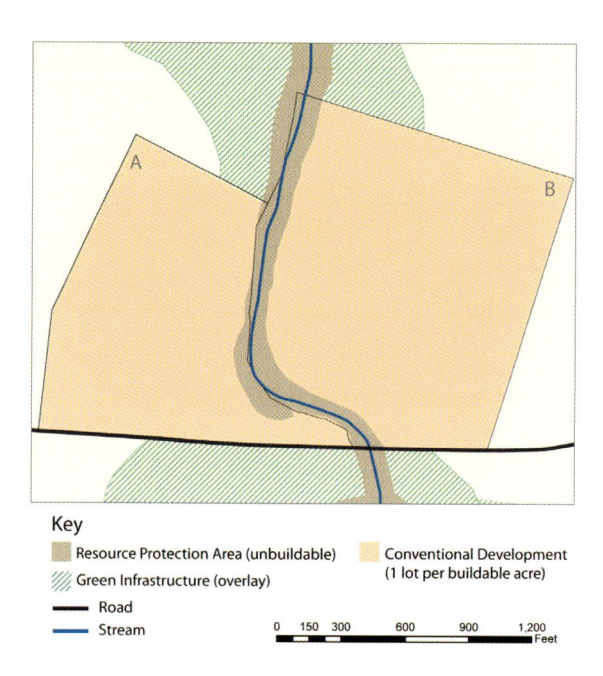

The habitat corridor had two parcels that could be built on (A and B); if fully developed, they would have narrowed the corridor and cores significantly, leaving only the mandated stream buffer. If that happened, the corridor would be too narrow to support extensive wildlife movement.

Green infrastructure: Map and plan the natural world with GIS

Key

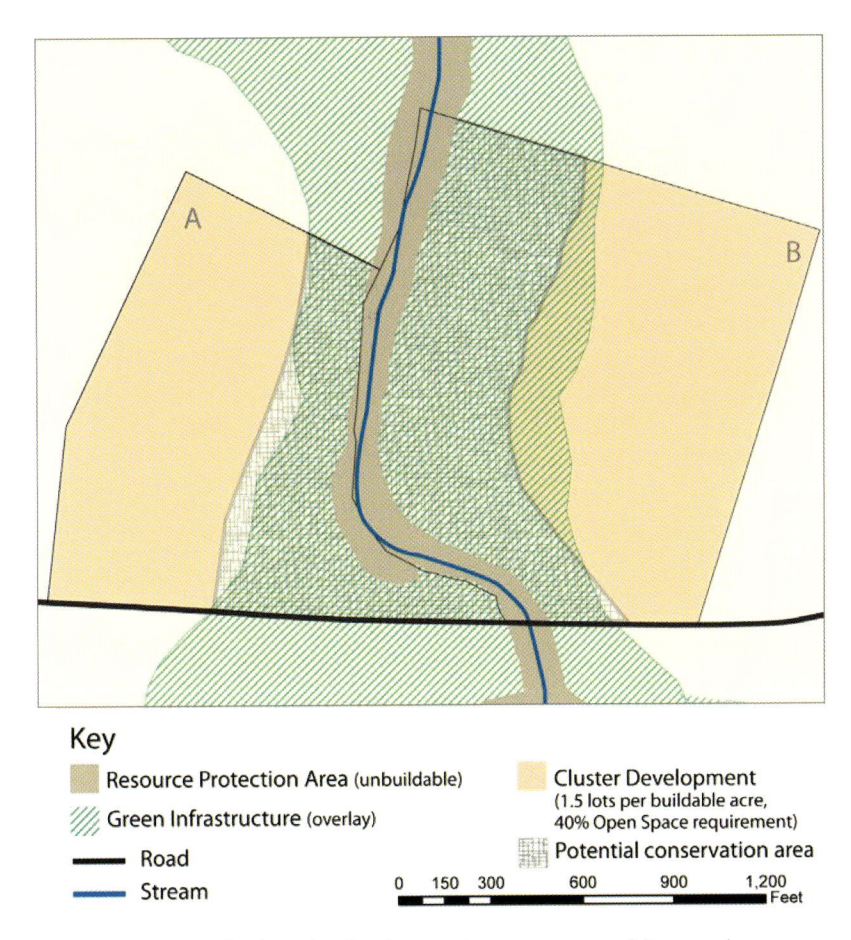

- ▓ Resource Protection Area (unbuildable)
- ▨ Green Infrastructure (overlay)
- ▬ Road
- ▬ Stream
- ▨ Cluster Development (1.5 lots per buildable acre, 40% Open Space requirement)
- ▦ Potential conservation area

0 150 300   600   900   1,200 Feet

By building on smaller lots, the developments can accommodate more houses and conserve most of the corridor. The pink areas are available to develop, losing only a small area of the corridor, whereas the double crosshatch area can be maintained. See chart below for the numbers of housing lots before and after clustering.

Development Parcel (acres)	# of Lots Conventional Development (1 lot acre)	# of Lots Conventional Development (1.5 lot acre)
A (32)	26	39
B (36)	30	45
Total	56	84

The left column for conventional development shows that a total of 56 lots can be constructed, whereas 84 lots can be constructed using a clustered development scheme building on smaller lots while conserving most of the habitat corridor. Credit: Green Infrastructure Center Inc.

As explained earlier, homes near green space generally sell faster and for higher prices than homes farther away. Therefore, this development would cause fewer impacts and also be more profitable. By doing this analysis with the help of local planners or landscape designers, GI plans can be shown to be realistic and doable.

Chapter 7 | Determining opportunities

Case Study: Adams Park scenario

Publishing core maps is the first step in letting landowners, planners, and designers know about the presence and importance of habitat cores. In the next example, a subdivision has been plotted but not built. With good planning, a developer can preserve the existing woods along an impaired reservoir. By redesigning the lots, the developer can conserve more land to protect the reservoir and core connections. In this case, the developer of this site built 12 additional homes by clustering while conserving open space.

Development scenarios

Four scenarios highlight different development and neighborhood amenity options on the basis of a higher-density development pattern. In the first image, existing development shows the forest core is not yet developed: in scenario 1, the developer's proposal is shown with most of the development completed and forest removed; in scenario 2, the same site has smaller lots, less roadway, and a nature trail; scenario 3 shows the same number of lots but with open space easements across the backs of the lots to allow more green space and expansion of the nature trail; finally, scenario 4 shows the development with added rain gardens and green fingers (strips of landscaped areas).

Some people may have wanted to protect the entire forested area as a nature preserve, but in urban areas, shrinking the development footprint from 30 acres to 10 is still a worthwhile outcome.

This example of scenario building is one way to help the development community envisage other ways of development. Little by little, these approaches can string forested corridors between and within developing areas.

Adams Park was permitted to develop all the land with no open space. The Green Infrastructure Center modeled scenarios first with smaller lots and open space, then smaller lots with trail easements at the back of the lot, and finally with smaller lots and adding bioswales and other low-impact development. Credit: Green Infrastructure Center Inc.

Pollution-loading reductions (from developing less land) were also modeled to show how the project could meet goals for protecting a currently impaired waterway.

Chapter 7 | Determining opportunities

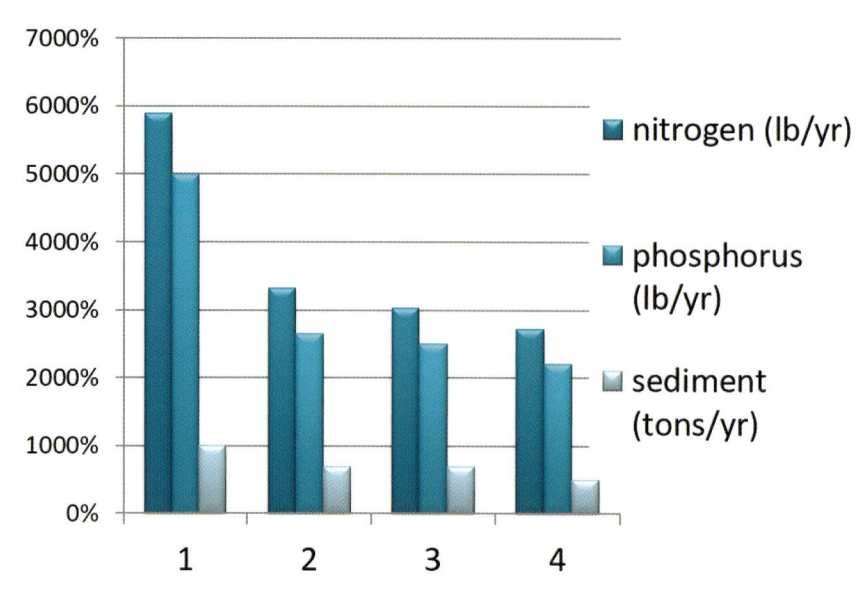

Pollution loadings fall significantly from the original design to the cluster design options. Credit: Green Infrastructure Center Inc.

Climate change

As we have seen, many risks translate into opportunities, and climate change is no different. With rising sea levels, areas along coasts that were once developed or extremely desirable for development have become more suited for open space, so this has become an opportunity. As noted earlier, you can model sea level rise (SLR) to an appropriate time frame depending on the planned duration of the project. That, in turn, depends on the goals. The planning horizon will be longer depending on the purpose (e.g., a longer-term goal would be to plan a road or new hotel, whereas a shorter-term plan would be a parking lot or kayak launch). The opportunity presented is to recognize and plan for new coastlines and the resultant changes in the locations of marshes, coastal forests, and recreation uses, such as boating access and fishing docks.

Hypothetically, when planning to plant a forest buffer to improve water quality, you may not want to plant trees in an area that will be underwater in a decade or so. Consider planting water-tolerant trees, such as bald cypresses or even mangroves, depending on the ecosystem. You can use GIS to determine future coastlines, marshland inundation, and locations for forest buffers. For naturally forested waterways, plan to keep them forested and replace forests where they have been lost. If rivers have other adjacent natural land covers (e.g., forest is not the climax land cover that would naturally occur there), then of course the buffer should be native vegetation. You might still decide to plant a forest buffer along today's shoreline to provide short-term habitat and a buffer from storms. But you also may plan for the future buffer in its new location, which could require purchasing land, changing the species of trees for a project, and procuring inland easements to allow the buffer room to migrate.

Changes to coastal GI should also include marsh migration. Marshes are part of GI and should be included in GI network maps along with rivers, lakes, and bays. The National Wetlands Inventory (NWI) has some of these data, but you should consult coastal scientists, university staff, and other organizations to determine whether they have more recent and specific data for the coastline and adjacent marshlands. As noted earlier, the National Oceanic and Atmospheric Administration is the usual source for US-based SLR data. Recall that SLR is not uniform along the world's coasts, because the ocean floor, currents, local water temperatures, and the makeup of the coast all affect SLR.

As the sea rises, water moves inland—where it can. With a high, steep coast, such as a cliff or sedimentary bluff, shallow marches may have nowhere to move. In such areas, shallow marshes may disappear, affecting the makeup of marine life that can forage, shelter, or nest there. But for other sections of coastline, where there are not steep elevations changes, marshes are likely to move inland. In such cases, consider where the new marsh will be located (hypothetically) and whether the marsh can form in its new location.

Barriers to the inland migration of marshes might include a marina, causeway, housing development; artificial berm; and other obstacles. With storm surges more likely, such barriers could become isolated if water pours over them during a storm and becomes trapped on the other side, thus forming a wetland that will not flush adequately and becomes stagnant or choked with invasive species, including grasses included in the genus *Phragmites*.

Evaluating Sea Level Rise
20 Year Scenario

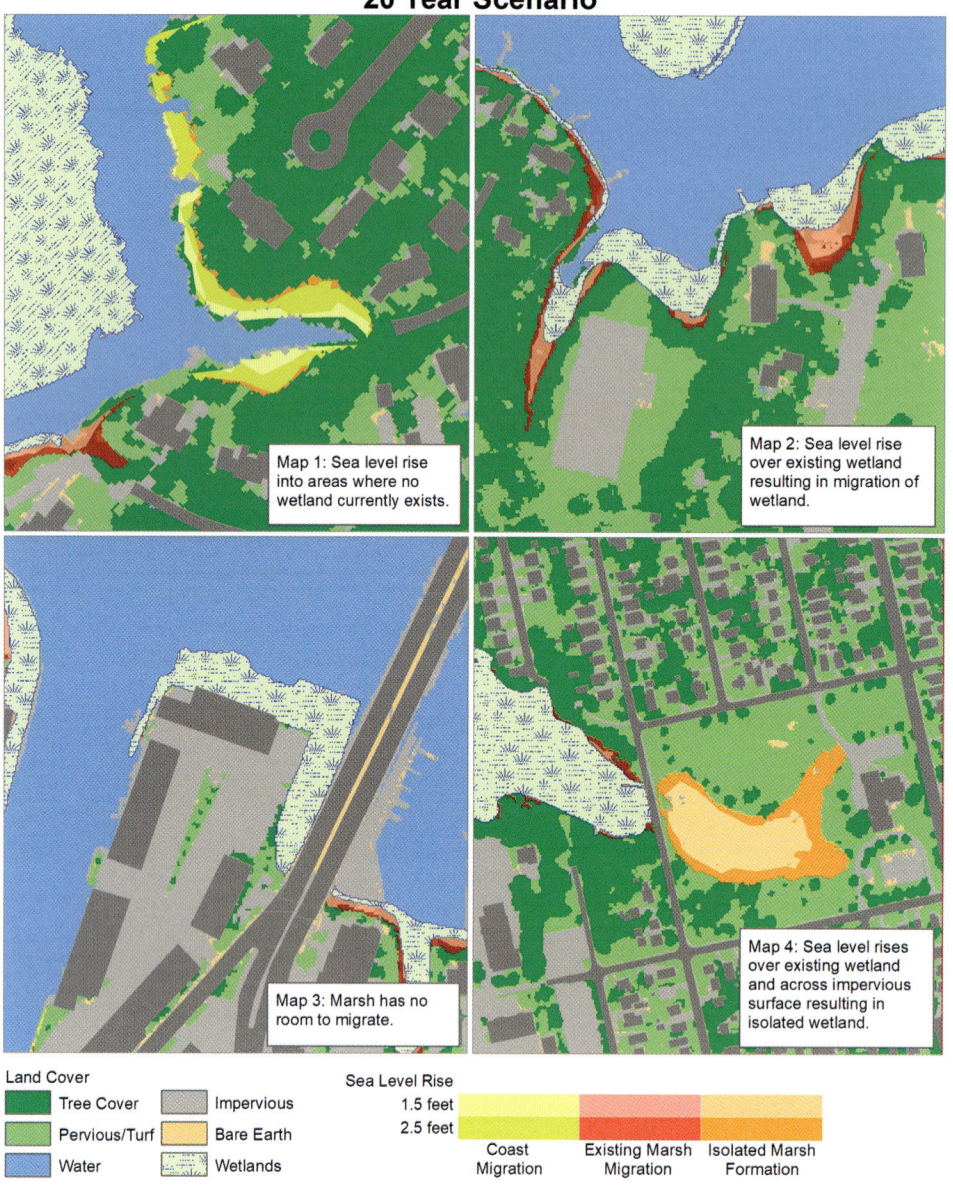

Map 1: Sea level rise into areas where no wetland currently exists.

Map 2: Sea level rise over existing wetland resulting in migration of wetland.

Map 3: Marsh has no room to migrate.

Map 4: Sea level rises over existing wetland and across impervious surface resulting in isolated wetland.

Land Cover: Tree Cover, Impervious, Pervious/Turf, Bare Earth, Water, Wetlands

Sea Level Rise: 1.5 feet, 2.5 feet — Coast Migration, Existing Marsh Migration, Isolated Marsh Formation

In this scenario, sea level rises ranging from 1.5 to 2.5 feet are modeled to see areas where the sea may rise and cause new marshes, map 1; areas where the sea will rise and cause an existing wetland to migrate inland, map 2; areas where the sea will rise but there is nowhere for marsh to form, map 3; and areas where the sea rise results in an isolated wetland forming inland, map 4. Credit: Green Infrastructure Center Inc.

In cases in which GIS analysis shows that the coastline has a lot of room to migrate, then there is nothing to do except to create an overlay so that planners, ecologists, and others can access the likely location of the future marsh, coastal dunes, or beaches.

The following list includes some of the questions you can ask to surface opportunities for action:

- Will the entire park and all its facilities be lost (and if so, when)? Should new areas be sought to replace the lost park?

- Is there an opportunity to use the current park as a wetter open space by installing boardwalks, viewing blinds, kayak launches, and other low-impact infrastructure, which will allow visitor access?

- If a marsh will be lost, should other marshland be protected to ensure that the same amount of coastal habitat is available in the future?

- Are other actions needed, such as restoration of inland areas, the relocation of housing developments, and removal of coastal roads, to allow the marsh to migrate? Other examples might include removing a bulkhead barrier, restoring a natural slope, digging out old fill to return fields to a natural marsh, and protecting an existing marsh with a notched breakwater sill to absorb wave action while allowing easy movement of marine life.

- Does the marsh or tidal creek serve as an important wildlife connector? Will any connections be lost with SLR? Do connections need to be reestablished with new plantings?

Multibenefit thinking

Sometimes, you can creatively meet multiple goals to foster new opportunities. One example is the use of oyster reefs to prevent waterborne terrorist attacks. This idea seems a little strange, even counterintuitive. However, landscape architects and ecologists who work for the US Navy realized that building up oyster reefs in waters around naval bases could be funded as an antiterrorism mechanism because the reefs would prevent even small craft from accessing some areas of the base. Similarly, removing invasive *Phragmites* with shorter marine grasses to restore habitat improves visibility around the bases, and thus improves safety. Other real examples include planting meadows instead of lawns to save funds on moving and improving habitat and infiltration of water.

Users should think creatively when looking for ways to justify projects. How many ecological, social, and financial goals could be met? Will replanting forest habitats attract more birders who could also bring tourism dollars as they stay overnight, buy meals, and shop? Will restricting floodplain development and restoring riparian areas prevent property losses and earn credit from the Federal Emergency Management Agency (FEMA) to lower insurance rates for coastal landowners? Could a wildlife corridor also become a greenway trail and attract more tourists and companies looking for green communities to locate their businesses? To learn more about GI, see the author's previous book, *Strategic Green Infrastructure Planning*. The more benefits that can be determined, the greater the likelihood that an opportunity will be recognized, appreciated, and, if needed, funded.

Climate change is not just about SLR. As discussed earlier, some areas of the globe will become warmer. Some species will adapt or move. For example, species such as the American pika may migrate to higher elevations to reach the habitat conditions it needs to survive, whereas other species, such as the peregrine falcon, can adapt by building nests on tall buildings as natural cliff habitats become less abundant. In other areas, smaller streams or even some large water bodies and rivers may dry up as lands become warmer and more arid. This type of risk may be harder to overcome because this habitat will disappear rather than move inland. In this situation, a core that still supports water because it has a perennial river may become more important for wildlife as smaller streams and pothole wetlands are lost. The plants that made up the core (trees, shrubs, tall grasses) also may change and the core itself could disappear.

However, other areas could become wetter as El Niño and other weather patterns change or storms become more severe. If these changes affect the GI plan, one implementation action could be to expand emergency planning or budget more funds for habitat restoration, such as replanting devastated areas and restocking fisheries.

As the Ersi GI model is updated with new land cover from National Land Cover Database imagery, these changes may become apparent. Areas that once supported agriculture may become too dry to farm effectively. Land-use changes could be dramatic over time. Species may disappear for a variety of reasons, including loss of access to water, changes in the timing or availability of food supplies, or inability to adapt to warmer temperatures. For the food web, warmer temperatures could cause fish to spawn sooner and not be present when large mammals arrive for fish runs in the spring. Migratory diadromous fish that live part of their life cycle in marine water and part in freshwater may also be affected when one or both of those environments are altered.

As areas become more sensitive to droughts, withdrawals of water may need to be reduced. For example, the western US follows the riparian water rights system (first in time, first in use) and allocates water uses based on historical permits. Water uses could also be affected by faraway desert cities where people water their lawns to mimic water-rich climates instead of acknowledging the desert climate in which they live.

These complex issues may be beyond the scope of a GI plan. However, understanding these "change drivers" and their possible outcomes will allow a plan to be more realistic and adaptive.

This chapter showed creative examples of ways to preserve cores and corridors, at least in part, even within a developing landscape. The chapter showed how to use GI maps to surface new needs and opportunities for conservation and enlist diverse interests to help recognize these strategic opportunities. The next and final chapter discusses implementation and focuses on several case examples of GI plans to provide motivation and inspiration. A GI plan is only as good as its implementation.

Notes

1. Brown, J. W., *Eco-logical: An Ecosystem Approach to Developing Infrastructure Projects*. No. DOT-VNTSC-FHWA-06-01 (Cambridge, MA: Research and Innovative Technology Administration, 2006).
2. Beier, Paul, Wayne Spencer, Robert F. Baldwin, and Brad McRae, "Toward best practices for developing regional connectivity maps." *Conservation Biology* 25, no. 5 (2011): 879–892.

Chapter 8

Implementing GI plans

The previous chapters covered the steps for creating a GI plan: setting goals, obtaining data, making maps, assessing risks, and identifying opportunities. The final, and most important step, is implementation.

Implementation of GI plans can take several forms and depends on the goals set in step 1. If, for example, a plan established a goal of fostering recreation by creating new trail connections, the risk phase may have identified areas that could be lost, and the opportunities phase then identified places where trails are desired. This final phase addresses how to develop effective implementation strategies—what to tackle first, how to do it, who does it, and what funding is needed.

The most important form of implementation is using the maps for everyday planning. GI should inform every aspect of planning for development, recreation, water supply and health, public safety, and conservation. Maps of GI should underpin all decisions regarding where to conserve land, where to develop, and how to develop in ways that allow for connectivity across the landscape.

Natural asset maps

The natural asset maps created for a GI assessment should form the basis for the following strategies:

- Planning development that keeps the landscape connected; determining where growth should and should not occur
- Reviewing site plans for impacts, avoidance of barriers, and restoration of green assets and connectivity
- Planning for long-range goals, such as comprehensive plans or economic development plans
- Updating the zoning map (changing old zoning, adding new restrictions, allowing and disallowing uses)
- Locating new parks or other open spaces (e.g., to protect rare species, restore habitat, add connectivity)
- Prioritizing which lands to target for conservation easements
- Determining the least impactful routes for new roads (or planning for alternatives to move people)
- Protecting the landscape context for historic and heritage sites
- Making the case to decision makers about conservation or growth-planning needs
- Planning for future water supply (e.g., source protection, intake protection, recharge protection)
- Planning for disasters and emergencies (e.g., what other areas are at risk and should not be developed, and which shoreline buffers could be restored?)
- Planning for new recreation uses and locations (e.g., birding, mountain biking, kayaking, and hiking)

Decision support: Using GI maps and apps for everyday planning

As noted at the beginning of the book, GI planning considers natural assets in everyday decision making. Now that data and maps have been developed, they should change the way planning is conducted, whether for a local or regional government, state agency, land trust, conservation group, or other civic or community groups.

The mapping group should include key players during the entire process, so that all decision makers involved in community planning are on board. However, if the planning has been internal to one department, now is the time to talk with other departments about ways to use the new data to better inform their planning. Planning with landscape connectivity in mind is not yet the norm in America—or indeed in most countries. This chapter will showcase several GI plans, including one plan where GI planning is mandatory in Spain.

Overcoming stovepipe planning

When a local government operates in a *stovepipe,* or *isolated,* fashion (as governments often do), one agency deals with site plans, another plans roads, another plans for parks and public open spaces, another plans for economic development, and so on. Working in isolation this way can present challenges when decision making requires integrated planning across disciplines. It's often difficult for one department to fit a trail into a development plan or reconfigure a site plan to protect a wetland or large forested core when other departments are unaware of the authority's environmental or recreation goals. Often, such considerations occur only late in the development process after significant time and money have already been expended and plans are fairly rigidly fixed. For example, realigning a road network already laid out to avoid a key core or protect a species could require considerable effort and additional cost.

To address these concerns early, and thus more cheaply and effectively, community authorities can change the way they operate. Key agencies can easily give up their disjointed approach of operating in isolation and collaborate by sending representatives to predevelopment meetings. This approach promotes cooperation and an exchange of ideas early in the process. In terms of GI planning, this approach allows agencies to raise concerns about species habitat, core areas, and connective corridors before plans are solidified, allowing for landscape connectivity.

Growing communities can invite developers early to discuss ideas, especially for those properties identified in the risk analysis as having abundant green assets and imminent development threats. For example, a community may find a specific property worth acquiring as a park or other public open space to ensure the connectivity of a local trail, protect a particularly beautiful stretch of river, or protect the community's drinking water. Communication and collaboration are essential because most of the land under consideration is likely to be under private ownership.

> A community in the southern US wanted to attract new companies to the area. The community used its GI map to create a competitive advantage over neighboring counties by showing that the community was green and progressive, with a healthful environment and plenty of access to natural areas—factors that many companies appreciate when considering where to expand in new territories.

Projects initiated by other public bodies

If a land trust, university, or other conservation group initiates a project, now is the time to talk with local decision makers, if that has not yet happened. As explained earlier in this book, the decisions of local, regional, and state governments greatly affect land development. GI planning requires that these entities participate in any project to map and conserve habitat cores and corridors.

Many large conservation groups, such as the Conservation Fund and the Nature Conservancy, long ago realized they could not purchase every key parcel of land or get conservation easements placed on every acre of a particularly significant landscape; cooperation and planning with local, regional, and state agencies are critical to the effectiveness of their missions.

Groups such as the Green Infrastructure Center help local governments assess and plan for GI conservation or restoration in concert with all levels of decision makers. Local land trusts and other regional organizations can also be key partners in helping local governments identify important landscapes for permanent conservation. In addition, many counties also hold easements to avoid developing key parcels and compensate landowners for giving up development rights. For example, a county or land trust could obtain an easement for parcels that protect water recharge areas for a community's water supply.

Using GI maps to save money!

Planning that happens well in advance usually saves money. Early planning allows projects to consider how best to link to existing resources and avoid the expenses of overclearing land or planting trees to replace those that had been removed. Popular amenities, such as regional greenways, water trails, and land trails, can be included in the plans. These plans also can integrate scenic vistas, improved transportation networks, and other forms of access to nature at an early stage.

Early GI involvement can substantially reduce development costs

Early GI planning also can reduce development costs by designating more land as open space (undeveloped) and thus reducing the volume of stormwater runoff requiring treatment. This saves money and increases profits. Furthermore, better informed and longer-range planning helps communities allocate lands for parks, schools, and trails. A GI map can also inform planners about which of those lands to acquire, connect, restore, or protect. The many available tools include fee simple purchase, conservation easements, zoning or zoning overlays, ordinances requiring open spaces and connectivity, actions of private landowners, the establishment of land preserves, and the use of land banks that hold land as mitigation for land disturbed elsewhere.

Informing strategic conservation plans

GI maps can demonstrate the worth of a project to potential investors. A GI plan shows that the community, land trust, or locality has a strategic process to determine what land it values, what to conserve, and how to conserve it. Many land trusts have used GI maps to create long-range strategic plans, as the following examples demonstrate.

Meeting regulatory needs

GI plans can inform many regulations and help meet incentives. With creativity, GI plans can also inform many conservation goals and mandated state planning requirements.

> ### GI planning in urban areas
>
> In urban areas such as towns and small cities, GI planning must incorporate thinking at smaller scales that consider relationships between the various ecological and cultural aspects of the urban environment. In cities, natural features such as springs, trails, parks, and historic neighborhoods make up a complex cultural and ecological landscape that depends on a healthy environment.
>
> At the urban scale, large habitat cores are not the primary focus, because most cities lack them (defined as 100-plus acres of intact habitat). As a result, smaller-scale habitat stepping stones (patches) are important as street trees, pocket parks, city streams, trails, lakes, and other open spaces. In cities, tree canopy is a primary GI focus. Tree canopy regulates temperatures, absorbs air pollutants, intercepts stormwater, and mitigates flooding. Tree canopy can be mapped by watershed to study the role of canopy in infiltrating and cleaning the stormwater.
>
> Even large cities can develop habitat corridors, especially along rivers, streams, and abandoned railroads, which can link to the environment beyond the city. Cities can connect these corridors to larger habitat cores on their borders, providing trails for animals, hikers, bicyclists, kayakers, and boaters.
>
> As a result, in urban settings, GI plans often create the following tools:
>
> - A high-resolution map of land cover to determine the extent of tree canopy cover
> - A map layer of cultural assets that can include historic structures, sites, districts, scenic vistas, and favorite places
> - A map of nature-based recreation features, such as bike lanes, trails, parks, overlooks, fishing spots, and boating launches
> - A map of high-value soils, urban farms, community gardens, and farmers' markets

National Flood Insurance Program's Community Rating System

Local governments voluntarily join the National Flood Insurance Program's (NFIP) Community Rating System (CRS) to earn flood insurance premium discounts for local policyholders. Governments receive points for actions or policies that reduce flooding and flood damage; these points earn premium discounts as high as 45 percent. The discounts help reduce costs, as flood insurance rates have skyrocketed in the past several years across the US.

The CRS has the following goals:

- Reduce flood damage to insurable property.
- Strengthen and support the insurance aspects of the NFIP.
- Encourage a comprehensive approach to floodplain management.

Communities can earn credit for community-adopted management plans that protect the critical natural functions of floodplains and native species in the floodplain, while implementing habitat restoration projects. The GI plan can meet certain requirements to earn points in the CRS program. The requirements include an inventory of all species in the plan's geographic purview, action items for protecting one or more of the identified species of interest and natural floodplain functions, and a plan review and update every 10 years.

Many GI strategies also receive credits through the CRS. A few examples include preserving land located in the floodplain as open space, preserving or restoring shorelines and channels in their natural state, stormwater management, regulating erosion and sediment, and including various aspects of flood-related information on maps and in GIS databases. The city of Norfolk, Virginia, used its urban GI Plan to apply for additional discounts under the CRS program.

State wildlife planning

All states in the US must develop wildlife action plans to identify species at risk and provide support for conservation action. The West Virginia Division of Natural Resources (WVDNR) created a conservation action plan to guide the protection of species diversity in the state. The plan aims to "conserve the diversity of West Virginia's fish and wildlife resources by emphasizing [those] species in greatest need of conservation." The plan identifies fish and wildlife resources that are Species of Greatest Conservation Need and proposes the conservation and management techniques for their preservation. To accomplish this goal, the state identified priority habitats to support them. As part of that task, the state developed a GI network map with help from the Conservation Fund. The state's land trusts also have used the data to identify conservation priorities and produce an interagency coordination tool to inform cross-agency planning for species conservation.

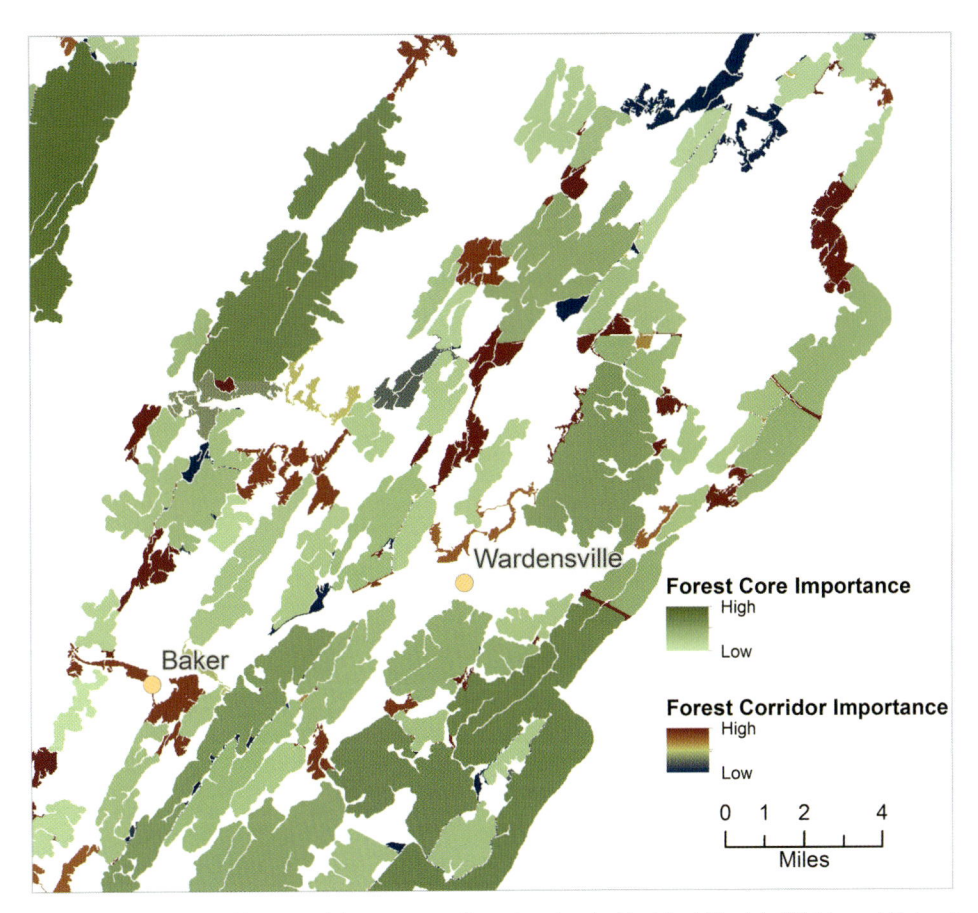

Forest cores and corridors used for conservation planning by the West Virginia Division of Natural Resources. Credit: Conservation Fund.

The following case studies present various types of GI plan implementation. Hundreds of such plans exist across the US and Europe. As mapping and GI planning become the norm, you will find them in nearly every community. See the Esri GI web page featuring a map of projects in the US: **https://urbanobservatory.maps.arcgis.com/apps/StoryMapCrowdsource/index.html?appid=9f39b46d4e40442f887ba80f6bc67772**.

Case studies

Case study 1: Informing land trust conservation priorities

Land trusts should have a strategic process to identify their conservation priorities. Although many conservation easements are opportunity driven—perhaps a local farmer places an easement on her land to avoid increasing tax burdens—land trusts also must ensure they are as effective as possible in meeting their goals. For example, in Virginia, the Capital Region Land Conservancy, which covers the counties and cities around Richmond, worked with the Green Infrastructure

Center and the Richmond Regional Planning District Commission (PDC) to evaluate high-value cores in the region. Creating a prioritized map allowed the conservancy to be more strategic when choosing which lands to conserve. The map also became a funding tool, showing potential funders that the conservancy had evaluated the landscape and was targeting easements for lands identified as having the highest value. The map helps the PDC assist localities with updating their comprehensive plans and looking for ways to encourage connectivity across borders at the regional scale. A series of workshops helps local governments identify priority connectors, especially connectors that cross boundaries and require coordinated planning to maintain.

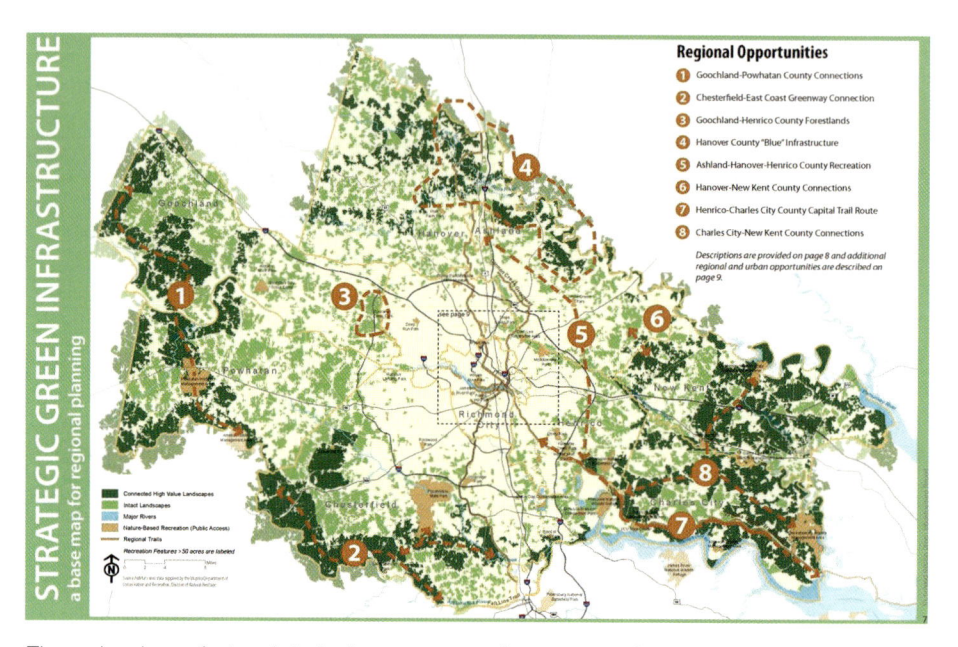

The regional map featured strategies across counties to protect key corridors. Credit: Green Infrastructure Center Inc.

In South Carolina, the Pee Dee Land Trust (PDLT) is using habitat cores to inform conservation priorities. Since its founding in 1999, the trust has conserved about 27,000 acres of land across 10 counties in South Carolina. Although using local staff and local knowledge has been effective in selecting areas for conservation easements, the PDLT wants to be as strategic as possible in the future by choosing the landscapes that meet its goals for conserving wildlife, water quality, and agricultural and historic resources. Using the habitat cores data and other data layers, the PDLT can inform its conservation vision to determine where to focus proactive efforts to achieve multiple benefits. It can also add protection for its existing conservation easements by choosing proximate, high-quality areas for new easements.

The PDLT used the South Carolina habitat cores and corridors model (developed by the Green Infrastructure Center, and upon which the Esri national model is based) as a base layer, on which they layered other data and models to select areas meeting multiple values. The final map identified thousands of acres of priority focus.

The PDLT used the Green Infrastructure Network Model for South Carolina as a base layer and overlaid that with the Resilience Network Model (RNM) developed by The Nature Conservancy. The RNM model selects species and areas that are better connected, and thus more resistant to climate change. Resilient areas are considered more resistant to climate change because they contain many connected microclimates and connect across the landscape, which tends to favor river corridors.[1] Data layers were incorporated for historic and cultural significance, water conservation priority, species richness, forested wetlands, and existing protected lands. Taken together, these data layers formed the final network. The PDLT found 431,246 acres of high-value habitats and 170,204 acres of high-value agricultural soils. These data will help the PDLT measure its progress in conserving the landscapes of highest value.

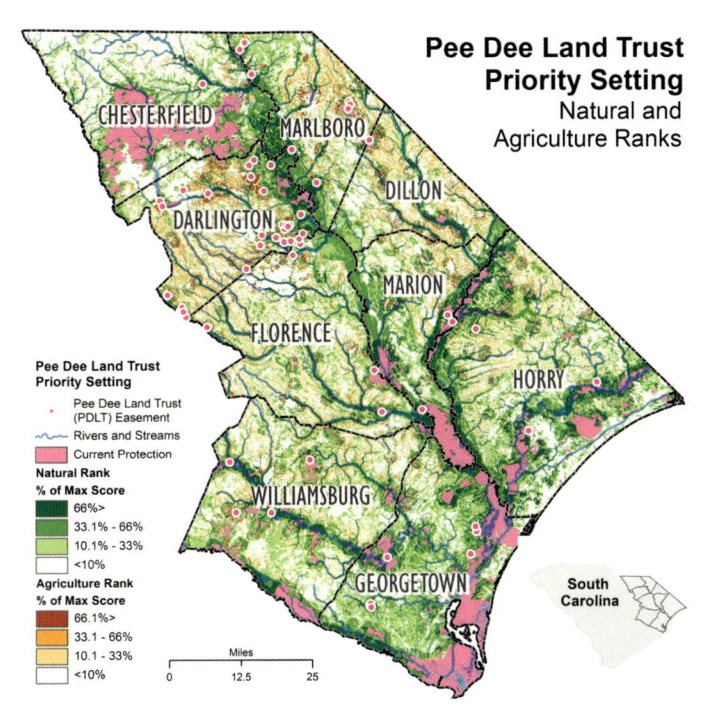

The Pee Dee Land Trust's green infrastructure priority map. Credit: Green Infrastructure Center Inc.

The PDLT uses its GI map to inform the focus of its conservation easement planning efforts. Places that hit multiple values will have priority in how the staff allocate its time when looking for suitable easements. The final map will serve as a spatial business plan, guiding the staff's focus for years to come.

The PDLT will accept easements from areas outside its mapped priority areas but hopes to accumulate or encourage easements in areas with the greatest ecological and cultural benefits. It hopes to remain resilient so that its investment meets its mission and it becomes more secure against potential climate changes.

Case Study 2: The Florida Wildlife Corridor

The Florida Ecological Greenways Network (FEGN), described earlier in this book, has led to some interesting offshoots. In addition to informing planning for the state's greenways network, the FEGN informs choices for spending and land acquisition for the Florida Forever fund, which to date has conserved 751,513 acres of high-value habitat. Through Florida Forever, the state has protected these areas:

- 604,930 acres of strategic habitat conservation areas
- 573,800 acres of rare species habitat conservation areas, including 899 sites that are habitats for 316 rare species (137 species are endangered, 56 species are threatened, and 20 species are of special concern under federal or state listings)
- 714,520 acres of ecological greenways
- 126,100 acres of underrepresented natural communities
- 523,420 acres of landscape-sized protection areas
- 384,440 acres of natural floodplains
- 726,580 acres that are important to significant water bodies
- 389,790 acres that will minimize damage from flooding
- 9,480 acres of fragile coastline
- 313,480 acres of functional wetlands
- 704,440 acres of significant groundwater recharge areas
- 425 miles of priority recreational trails
- 378,460 acres of sustainable forest land
- 1,019 archaeological/historic sites
- 11,970 acres in urban service areas

For more about the fund, see https://floridadep.gov/lands/environmental-services/content/florida-forever.

Another offshoot from FEGN is the Florida Wildlife Corridor. The FEGN's highest priorities, called critical linkages, identify areas needed to connect existing public lands with privately owned natural and agricultural areas for a functional, statewide ecological wildlife network. These linkages became the backbone for the Florida Wildlife Corridor.

The Legacy Institute for Nature & Culture (LINC), created as a nonprofit in 2007, manages the corridor. The LINC champions the permanent protection and connection of the corridor. It promotes the establishment and protection of the remaining natural lands, waters, working farms, forests, and ranches from the Everglades to Georgia and Alabama, protecting a functional ecological corridor for the health of people, wildlife, and watersheds.

The Florida Wildlife Corridor Expedition map shows the key wildlife connections across the state. Credit: Original painting by Mike Reagen, designed by Carlton Ward Jr., Tom Hoctor, Richard Hilsenbeck, Mallory Lykes Dimmitt, and Joe Guthrie.

The corridor came about in 2010 when partners throughout the state branded the FEGN's priority 1 and priority 2 lands and waters as the Florida Wildlife Corridor. The corridor was formally established by representatives of the Florida chapter of The Nature Conservancy, Conservation Trust for Florida, 1000 Friends of Florida, University of Florida, Audubon of Florida, Archbold Biological Station, Defenders of Wildlife, Florida Wildlife Federation, and the Florida Fish and Wildlife Conservation Commission. These conservation partners use the corridor map to inform their efforts to protect lands and waters through acquisition, conservation easements, and other mechanisms.

To elevate the importance of connectivity for wildlife, the organization has led two statewide expeditions traversing the corridor. This bold idea was born when Carlton Ward Jr., conservation photographer and organization founder, saw research by Joe Guthrie, a master's degree student at the University of Kentucky, tracking Florida black bears with GPS-enabled collars. Ward and Guthrie, seeing that the habitat range of one bear covered 500 miles over 2 months, proposed that a trio of expeditioners could take a similar path through the network of public and private lands. This trek was seen as a way to elevate the need to conserve natural and agricultural areas for wildlife.

After much planning, in 2012, the first expedition trekked 1,000 miles in 100 days to explore the wilds of Florida from the southern tip of Everglades National Park to Okefenokee National Wildlife Refuge in southern Georgia. In 2015, the second expedition, called the Glades to Gulf Expedition, traversed a 1,000-mile leg of the corridor from the Everglades headwaters in Central Florida to the Gulf Islands National Seashore in Alabama, passing through the Big Bend region and across the panhandle.

During the journey, the trekkers shared stories of ecological significance about longleaf pine restoration and Gulf Coast fisheries and the desirability of extending the corridor into Georgia and Alabama. "Trail mixers" held along the way allowed the public to join part of the journey. A book, video, and blog of the expeditions were created to record the scenery of natural Florida. Although the treks themselves did not conserve land, they drew attention to the need to connect, protect, and restore a network of lands and waters before they are lost. Work is under way to develop additional projects to protect the corridor's linkages and inspire citizens and partners to advocate for it.

Case Study 3: Urban GI in Hot Springs, Arkansas

Urban GI plans often encompass smaller-scale habitats, the overall tree canopy, springs and streams, greenways, and cultural sites. In the City of Hot Springs, Arkansas, one project conducted with the Green Infrastructure Center highlighted the variety and importance of natural and cultural elements in urban landscapes. Located in the Ouachita Mountains, the city is surrounded on two sides by Hot Springs National Park, which also owns and manages the area's historic Victorian Hot Springs bathhouses.

The city was founded because of the hot springs, which were originally enjoyed by Native Americans and trappers. Following the Louisiana Purchase in 1803, the land encompassing Hot Springs became a US territory. In 1807, the first settlers used the springs for their own health, and by the 1830s, cabins and a primitive store served increasing numbers of visitors. Many land disputes arose over access to the springs, and in 1832, President Andrew Jackson designated the area as the first federal reservation. During the following decade, the area transitioned from a frontier town to a Victorian-style spa city with grand, ornate bathhouses, the last of which was completed in 1888. In 1916, Congress established the National Park Service, and 5 years later the Hot Springs Reservation became Hot Springs National Park, enveloping what is today the City of Hot Springs.

To work on a GI plan, the city formed a Green Team made up of representatives from departments that affect land planning and management, including planning, public works, parks and trails, and economic development; the team included the city's board of directors and its city manager. Consultations with other key groups, including the Hot Springs National Park, historic preservation groups, and community members were also conducted. To reach people who might not otherwise attend meetings, booths were set up at the farmers' market to capture comments on the maps and gauge interest in key strategies. Tactics to enlist public interest included a sign that read Comments for Cookies, which enticed residents to engage with the maps and then get a cookie baked by the GIS team as their reward.

GI maps and plans should also note unique local geology such as the Arkansas Novaculite—uplifted hard rocks which led to naming the range as the "Zig-Zag Mountains," from which the hot springs originate. Credit: Green Infrastructure Center Inc.

Land cover classification

The high-resolution land cover dataset for Hot Springs was derived from National Agriculture Imagery Program (NAIP) 4-band aerial imagery. NAIP imagery is available at a 1-meter resolution, which is granular enough to adequately identify tree canopy. The imagery is taken when trees have foliage for easier identification of vegetation. This analysis reveals areas where water can soak in and be filtered versus areas where water runs off and carries pollutants and stormwater that impairs or floods streams and lakes. Natural resources data or assets show where the city's landscape supports good air and water quality, better shade, strong real estate values, and recreation. The data also identify landscapes and related amenities that may be deficient and need improvement.

The imagery was created using a supervised classification approach. Specifically, Feature Analyst™ software by Textron Systems was used (which can operate as an extension of ArcGIS) to identify land cover in the source imagery. The software can be "trained" to recognize the spectral characteristics of the land cover types of interest. The software classifies each pixel in the source image based on this spectral "signature" and the characteristics of the surrounding pixels. For better distinction between classes, a normalized difference vegetation index (NDVI) raster was created for use as an additional parameter in the classification.

Postprocessing

The raw classifications from Feature Analyst then went through a series of postprocessing operations. Planimetric data were also used at this point to improve classification. Roads, sidewalks, and trails were "burned in" to the raw classification. This step required the conversion of vector data to raster data, which then replaced the cell values in the raw classification.

The "tree canopy" class was not affected by the burn-in process because tree canopy can overhang streets. These data layers were also used to make logic-based assumptions to improve the accuracy of the classification. For example, if a pixel was classified as "tree canopy" but overlapped with the "roads" layer, it was converted to "Tree Cover Over Impervious."

The final step was a manual check of the classification. Several ArcGIS tools were custom built to automate this process. For example, the ability to draw a circle on the map and have all pixels that were originally classified as "tree canopy" be converted to "non-tree vegetation," which is a process usually requiring several steps, is now only a single step.

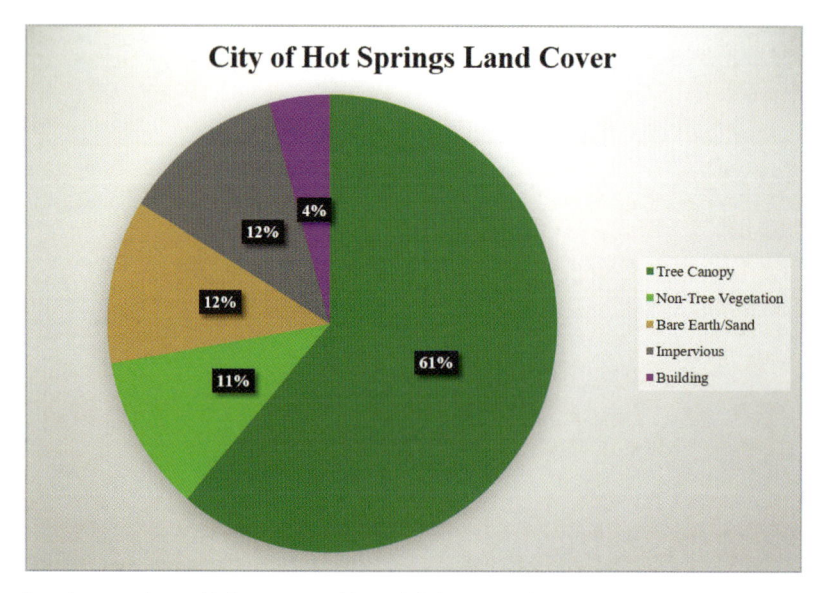

City of Hot Springs Land Cover

- Tree Canopy: 61%
- Non-Tree Vegetation: 11%
- Bare Earth/Sand: 12%
- Impervious: 12%
- Building: 4%

Land cover shows high canopy, although it is not uniform citywide. Credit: Green Infrastructure Center Inc.

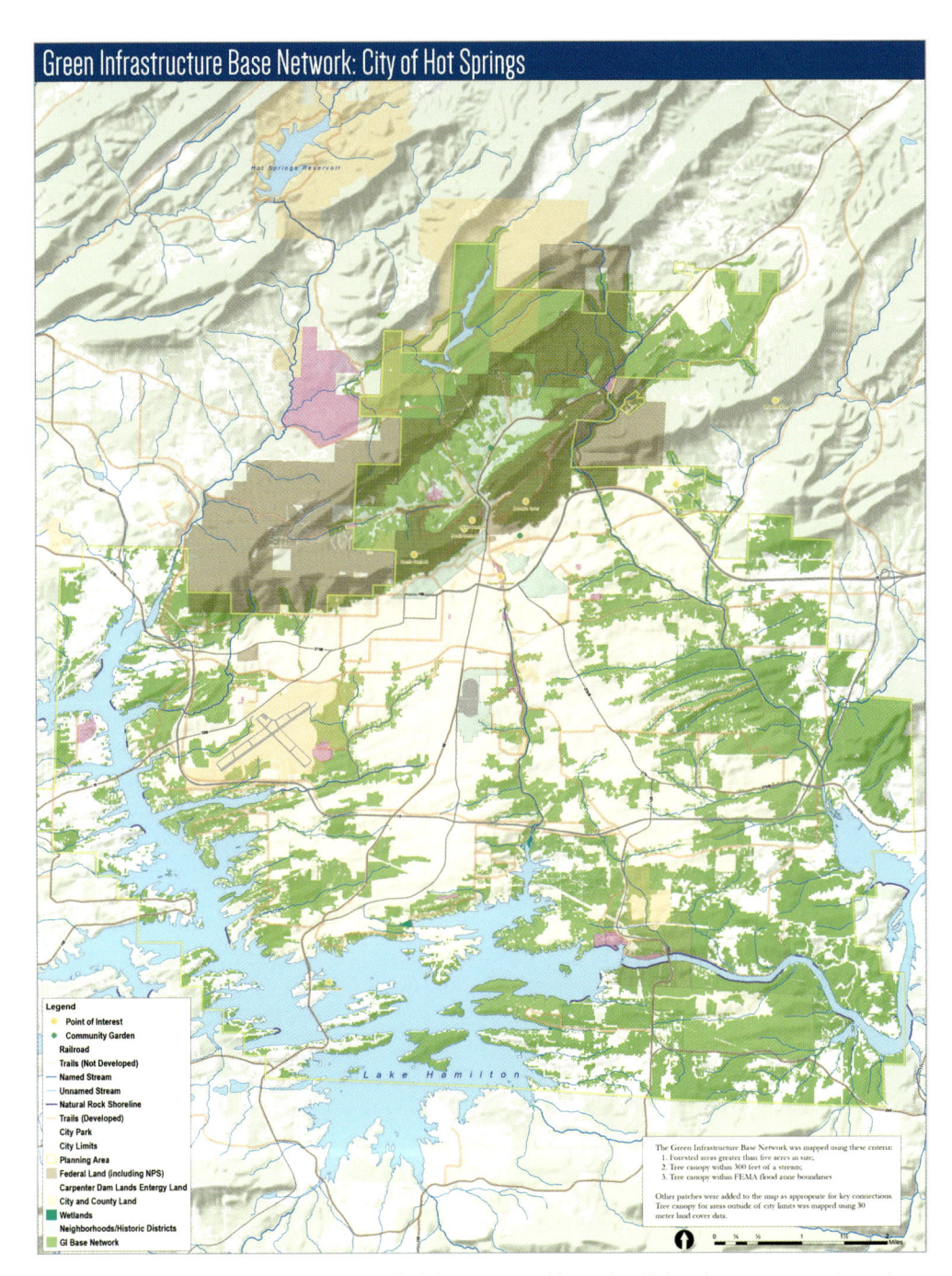

Green Infrastructure Base Network: City of Hot Springs

The city also considered the locations of habitat cores and how they link to key resources throughout the city and beyond. Credit: Green Infrastructure Center Inc.

The city studied the extent and location of these resources and then developed strategies to protect, restore, expand, and connect them. For example, the city prioritized the health of Hot Springs Creek, addressing flooding and reducing runoff into the waterway. Tree canopy is key because it intercepts, filters, stores, and transpires between 1.2 and 1.5 million gallons of stormwater annually, which equates to saving between $7 million and $9 million in stormwater treatment costs.

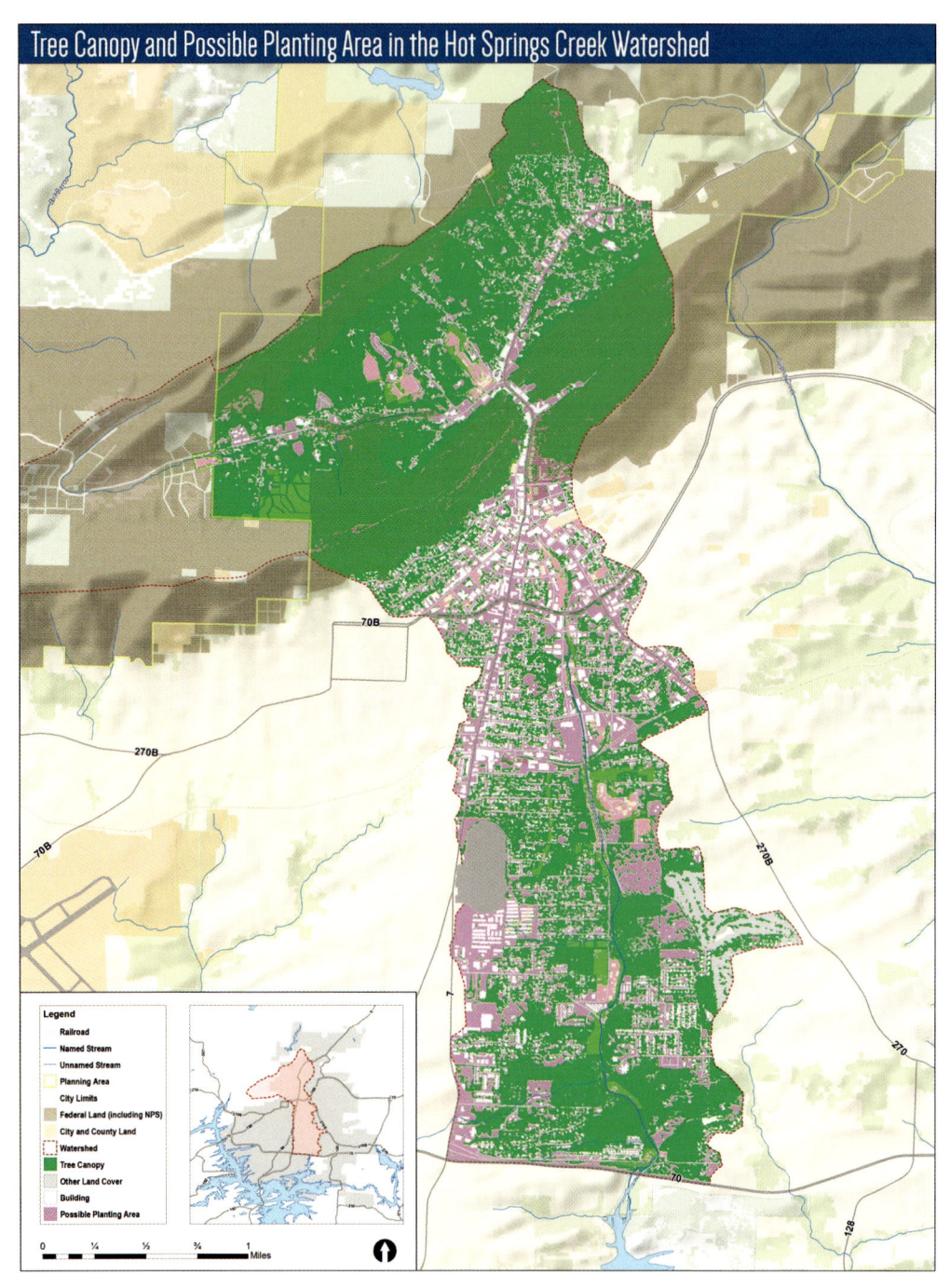

A map of existing canopy and also possible planting areas shows where the city can add trees in the Hot Springs Watershed. Credit: Green Infrastructure Center Inc.

The city's regional context sits in the path of a major monarch butterfly migration route and the migration routes of several other species. The monarch butterfly has declined to historic low levels, partly because of habitat loss along its migration routes. The maps helped the public see the city's place in the context of those routes and the national importance of the maps. Adding maps and context for species of interest can elicit more public interest and engagement.

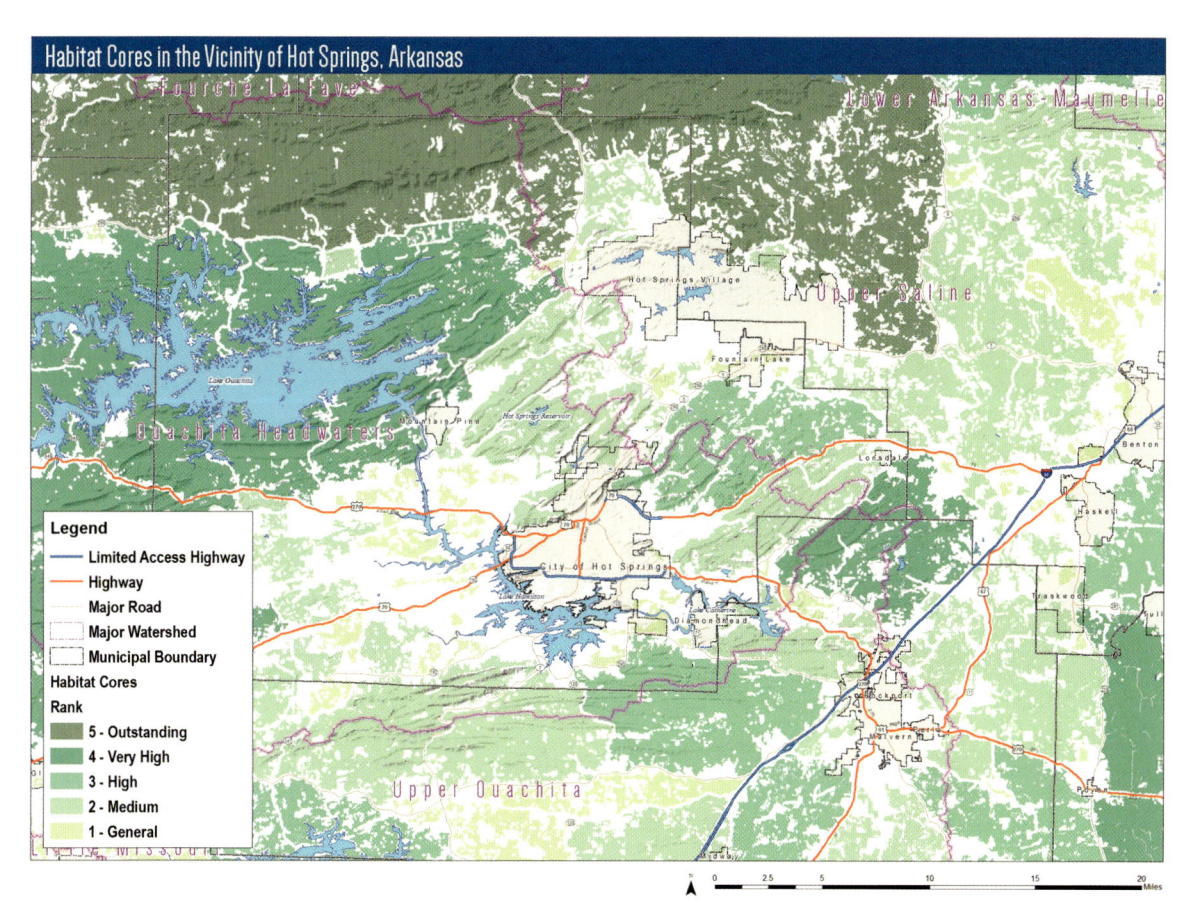

A regional map shows the relationship of the city to the cores habitat network. Credit: Green Infrastructure Center Inc.

Linking GI plans to the economy is another critical way to enlist public support. A healthy city is more attractive to residents and prospective employers, many studies have shown. Through its GI planning process, the Green Team identified many ways to increase opportunities to access the city's green assets, such as completing the Hot Springs Creek Greenway Trail, adding a water trail, and developing more forest trails in an area north of the city known as the Northwoods around the Lakeside Water Treatment Plant. The city also uses data for everyday planning and to inform long-range strategies for stormwater management, economic development, habitat enhancement, and public art.

The following identifies some of the city's strategies for its GI plan. You can find the full plan on the city's website under "GI" or linked to from the Green Infrastructure Center's web page (see Bibliography).

Hot Springs example strategies

- **Goal A: Connect the landscape to support wildlife, birds, pollinators, and people.**
  - **Objective:** Connect forests across the landscape—plant corridors to support wildlife.
  - **Objective:** Create a "neighborhood backyards birds, bees, and butterflies" program—the 3Bs!

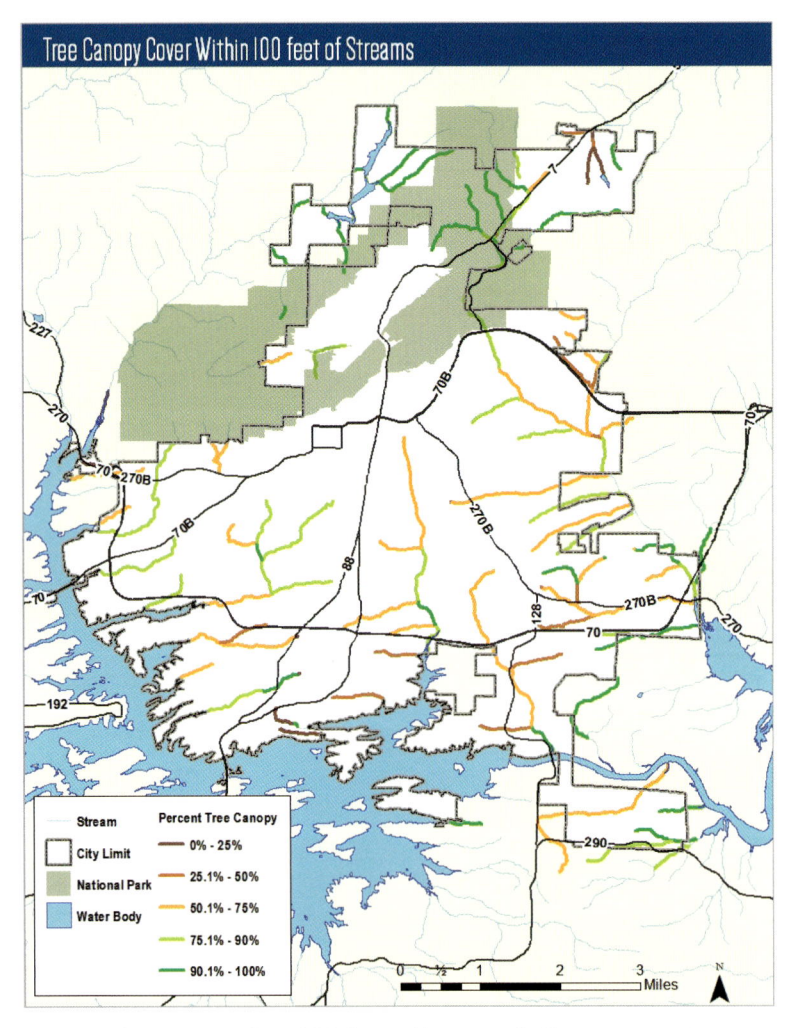

This map shows areas where 100-foot-wide stream buffers are present or needed to inform future planting priorities. Land cover is clipped for 100 feet on both sides of the stream and the percentage of coverage of trees is then calculated. Credit: Green Infrastructure Center Inc.

Green infrastructure: Map and plan the natural world with GIS

- Goal B: Support healthy creeks and reduce flooding by reducing stormwater runoff.

    - **Objective:** Replant areas along streams to buffer runoff.

    - **Objective:** Apply for grants to add permeable technologies, such as porous pavement or rain gardens.

- Goal C: Support economic health by protecting cultural and civic resources.

    - **Objective:** Improve community appearance and safety.

    - **Objective:** Support neighborhood-generated revitalization efforts.

- Goal D: Support healthy lifestyles by improving livability and tourism.

    - **Objective:** Create a paddling trails infrastructure to allow nonmotorized boating.

    - **Objective:** Develop the Northwoods Urban Forest Park trail system.

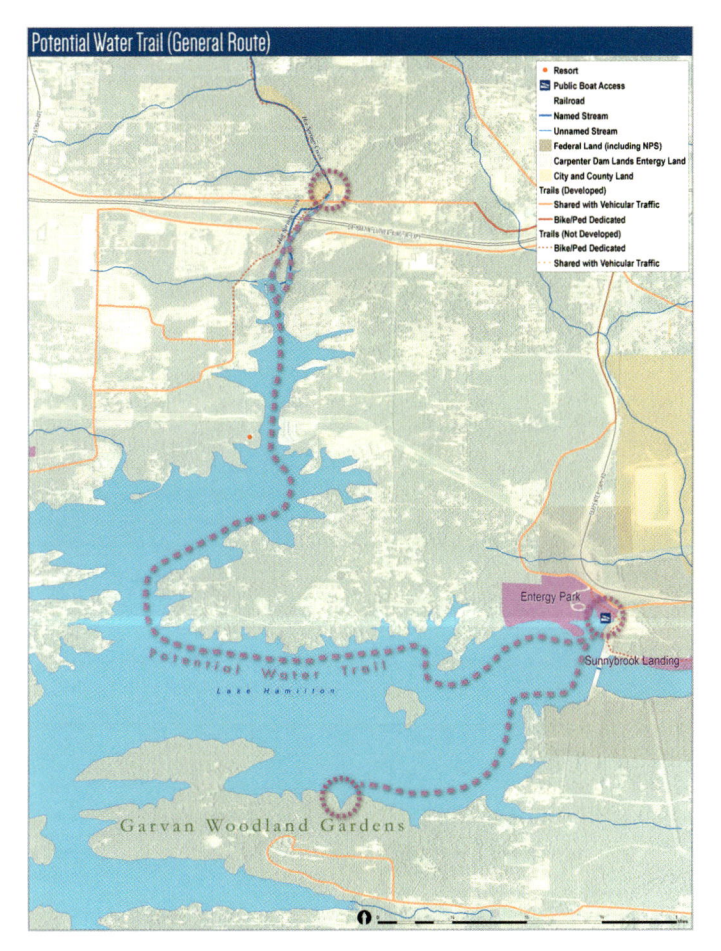

Future plans are to connect the greenway to a water trail. Credit: Green Infrastructure Center Inc.

Because the plan was adopted, the International Mountain Bicycling Association has constructed 15 miles of biking trails in the Northwoods area. This allows residents to access a far greater area of the habitat cores north of (and owned by) the city. These professionally constructed, world class trails avoid sensitive areas, minimize erosion impacts, and attract mountain bikers from across the nation.

To meet its GI goals to enhance access to nature, increase walkability, and treat stormwater downtown, the city created a detailed landscape plan for the Malvern Avenue Gateway, including a wildlife loop trail along sections of Hot Springs Creek, more public art to the streetscape, walkways and vegetation along the street, and a mile of bioswales to capture rainwater.[2] To download the plan's renderings, see **https://www.cityhs.net/659/Malvern-Gateway-Proposal**.

City staff meet twice monthly to ensure the corridor plan is on track and to identify new opportunities, such as retrofitting a parking lot along Malvern Avenue to add a pocket park. Interspersing green features along the road adds areas to infiltrate water and opportunities for the community to gather, adding to the city's livability.

This example shows that a GI plan is only a first step to increasing economic and ecological livability and that cultural goals can serve as a catalyst for other planning efforts. "Implementation" means using the plan to inform a wide range of strategies throughout the entire planning process, every day of every year.[3]

Case Study 4: The European Green Belt initiative

The European Green Belt initiative began in 2002 as a grassroots initiative to connect, protect, and restore habitat in a corridor running for 12,500 kilometers (7,767 miles) through 24 nation-states and encompassing eight biogeographic regions. Its vision is the following: "The European Green Belt, our shared natural heritage along the line of the former Iron Curtain, is to be conserved and restored as an ecological network connecting high-value natural and cultural landscapes while respecting the economic, social and cultural needs of local communities." **https://www.europeangreenbelt.org/**

Traversing 24 nation-states, the European Green Belt has provided a catalyst for conservation between and across countries. Credit: European Green Belt Initiative/Coordination Group.

The initiative has helped establish new parks and reserves, such as a biosphere reserve of 6,310 square kilometers along the Drava River between Hungary and Croatia. The initiative has increased the focus on key species, such as a 2005 project to protect the Balkan lynx (*Lynx lynx balcanicus*). In 2006, partners from eight countries undertook a gap analysis to evaluate the greenbelt as an ecological corridor. In June 2014, the initiative was formalized as the European Green Belt Association e.V., and had included 20 partners (countries and NGOs) by June 2016.[4]

A key aspect of the initiative is its dual focus on nature and culture. In the Saxony-Anhalt state in Germany, 92 hectares (about 230 acres) of open land were recently purchased north of the town of Salzwedel, protecting interior salt meadows and wetlands, as well as critical habitat for the Eurasian otter (*Lutra lutra*). The border strip that once divided East and West Germany now serves as a refuge for more than 12,000 rare and endangered plants and animals. A next step is to imagine the greenbelt as a spine for a more dispersed GI network reaching laterally across the countries it traverses.

GI planning is now a specific focus for the European Union. In June 2013, the European Parliament issued a communication, *Green Infrastructure (GI)—Enhancing Europe's Natural*

Capital, which calls for a standard part of spatial planning and territorial development that is fully integrated into the implementation of these policies, securing the resilience and vitality of some of Europe's most iconic ecosystems, with consequential social and economic benefits. The overall objective of the EU's GI policy is to have an EU network of GI for delivering essential ecosystem services throughout Europe.[5]

As these examples illustrate, GI planning is rapidly taking hold in multiple countries. Many of these plans engage stakeholders and include a process to incorporate broader social values, such as scenic views and culturally significant sites.

Exploring a green infrastructure network for Australia

Australians have been thinking about how to formulate a national GI network and have engaged in GI planning at the local scale for decades. Conservation scientist Robert L. Pressey of James Cook University has researched and written extensively on the need for more systematic conservation planning and recommended designs for nature reserves.

As recently as 2015, the Australian Institute of Landscape Architects called on Australia to lead the world in GI planning for a healthier nation, citing rising obesity as a key reason to plan for more green spaces and get Australians out and walking more. Currently protected areas are unevenly distributed across Australia. Australia is a signatory nation to the international Convention on Biological Diversity (CBD), which calls for the preservation of at least 10 percent of each ecoregion in the world.

Australia has a number of policy statements that support a GI approach, such as its *Biodiversity Conservation Strategy 2010–2020* (Natural Resource Management Ministerial Council 2010), the National Wildlife Corridors Plan (Commonwealth of Australia 2012), and special confirmation at the CBD's Tenth Meeting of the Conference of the Parties to the Convention on Biological Diversity (United Nations 2010).[6] Given the tremendous power of GIS in the 21st century, the time is ripe for Australia to create its first continental GI network model.

Case Study 5: GI planning leadership in the autonomous region of Valencia, Spain

Unlike most of the world, the autonomous region of Valencia (Spain) requires a GI plan to be included as part of *all* planning developments. Green infrastructure is defined by Spain's Law (17) as:

> the basic interconnected network of land composed of the following areas: the areas and sites of most relevant environmental, cultural, agricultural, and landscape values, the critical areas of the territory whose transformation involves environmental risks or costs to the community, and the territorial network of ecological corridors and functional connections for linking all the above.[7]

As a result, the region's 550 municipalities are legally required to create a GI plan before they implement any of their general plans.

In 2010, Arancha Muñoz-Criado, general secretary for Urban Planning, Landscape and Environment of the autonomous region of Valencia in Spain, led the effort to create Europe's first legal planning framework based on GI. Her passion for landscape planning was inspired by the changes she saw in her own landscape as a result of rapid, uncoordinated development. She was also profoundly inspired by Frederick Law Olmsted's interest in a connected green landscape as a primary construct for landscape planning.

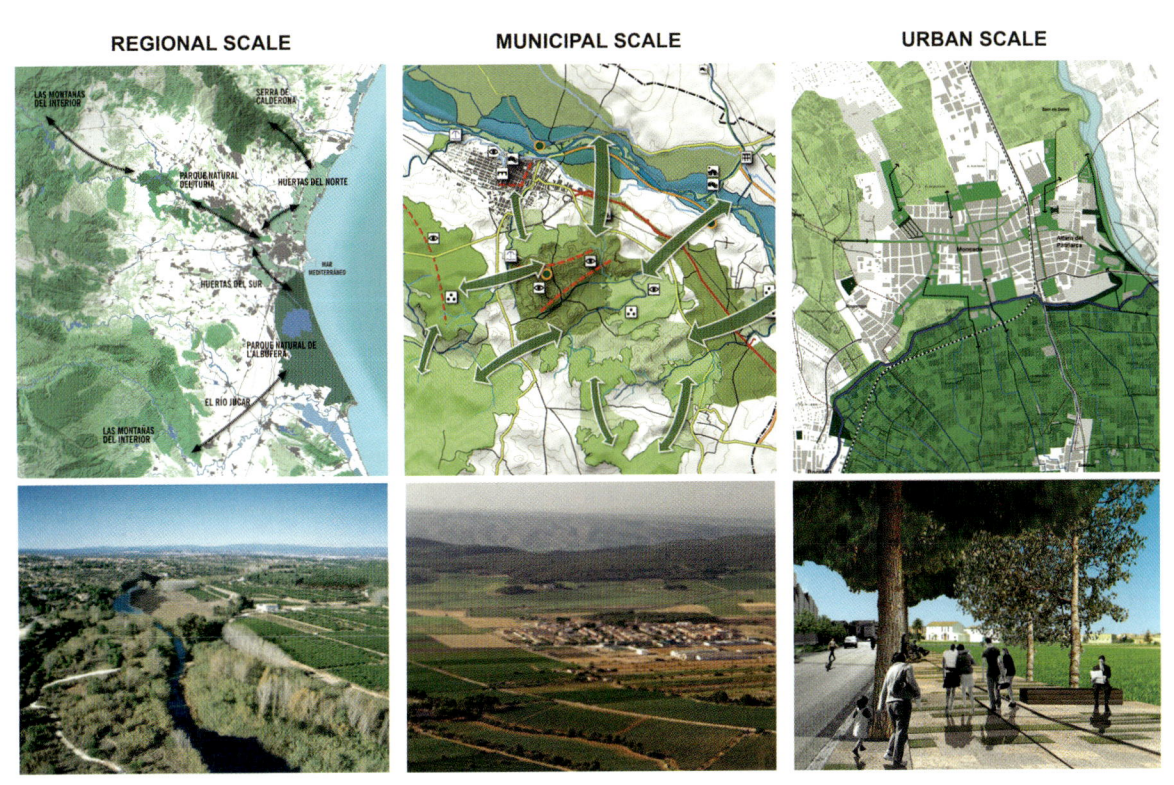

Green Infrastructure Framework for the Autonomous Region of Valencia, Spain

Green infrastructure at all scales. Credit: Arancha Muñoz-Criado.

Principles and criteria for the functional maintenance of GI in Valencia are set out in the region's Strategy Plan for the Region of Valencia 2030, the Green Infrastructure Regional Plan, and a new legal framework.[8] The regional government reviews and approves all development plans to ensure that GI is a key component and that the plans comport with identified regional GI goals and networks. Additional restrictions may further conserve specific GI resources.[9] To show how to implement this work at multiple scales, Muñoz-Criado led the development of plans and projects from the regional to the local level.

Chapter 8 | Implementing GI plans

The 2011, the Green Infrastructure Regional Plan set a goal to create an "ecological and functional network" to shape its future development.[10] The regional plan encompassed several strategies to protect or expand the region's GI:

- Identify and connect the most valuable landscapes with green corridors.

- Integrate urban connections.

- Improve visibility from the urban areas into the scenic rural landscape.

- Develop agricultural policies to protect the region's agricultural uses and heritage while affording greater public access and awareness of the area as a means to expand public support for its protection.

Water formed the basis of the Valencia Regional GI plan (Strategic Regional Plan for the Region of Valencia 2030), which included a series of subregional landscape plans that encompassed complex or sensitive areas connected by water.[11] Subregional plans include the delineation of local urban GI features, such as river corridors and walking paths. For example, in the metro area of Valencia, the regional River Turia Park connected to the sea through ecological corridors that end in oceanfront parks.

The region's GI policy aims not only to restore the landscape's environmental functions, but also to improve the visual quality of the region and the public's access and enjoyment. A photograph-based survey involving 800 people determined the desired visual qualities.[12] The resulting Visual Landscape Study of features ranged in a spectrum from "most preferred" to "least preferred." Based on these prioritizations, guidelines were created for governments and developers.[13] Today, all municipalities and developers must prepare a visual and landscape study that incorporates the most valuable views and landscape elements into the GI plan.

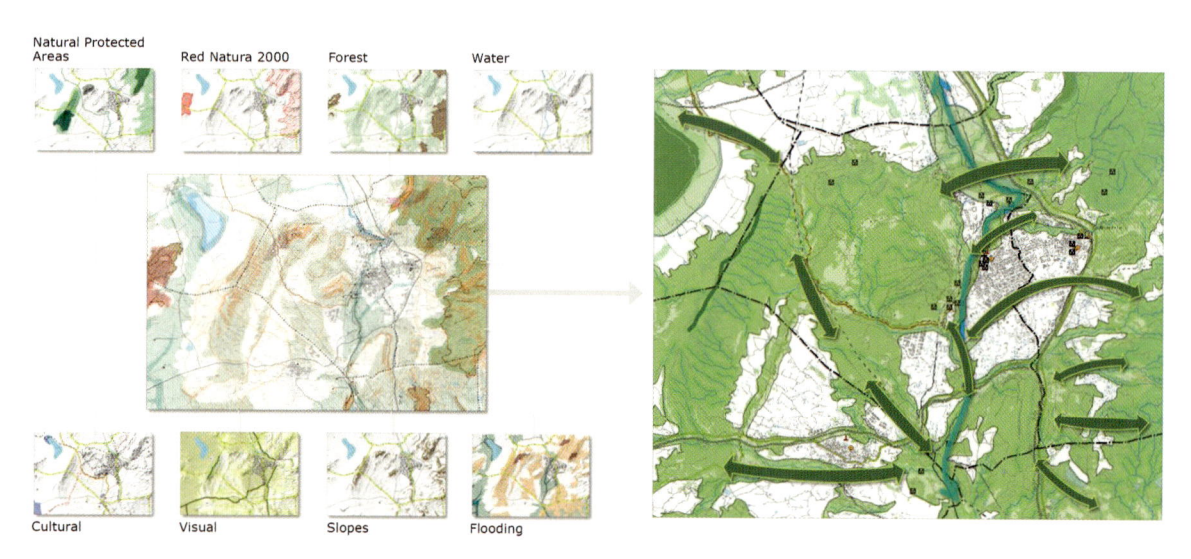

Green infrastructure layers and connections. Credit: Arancha Muñoz-Criado.

In addition, a GI plan was developed for metropolitan Valencia, which has a population of around 1.6 million people organized into 45 local municipalities.[14] The metropolitan area consists of four major high-quality landscapes that share water as their primary focus. The area has a millenary irrigation system of water channels dating to the 10th century. To the west is the Turia River Natural Park, with its Mediterranean and riparian forests; to the south, the Albufera of Valencia is a large wetland separated from the sea by a sandbar; to the east, the Mediterranean Sea has extensive sand beaches and dunes; and to the north, the Huerta de Valencia is an irrigated agricultural and cultural landscape that surrounds the city of Valencia and its neighboring towns and villages.

The area GI plan aimed to protect these four major landscapes and integrate them into an ecological and functional network. For the metro area, the plan focused on GI and the integration of regional and local plans. It has led to more bike paths, pedestrian walkways, green spaces, parks, gardens, viewshed protection, and habitat protection. Because these policies also enhance the regional agriculture, protecting much of it from development, residents now appreciate the connection between a strong agricultural region and access to local, healthy food and supporting cultural landscapes. Developers know which areas to avoid and how to protect areas they develop near key GI resources.

Conclusion

This book has outlined the science of GI planning and demonstrated how its processes and tools can be integrated into a wide range of planning scenarios. It has identified the central elements of GI, such as habitat core and connecting corridors; how they are defined, measured, and evaluated; and how a network of green assets can be mapped by a community and used in a variety of ways to enhance the goals of that community and its strategic aims.

Through six defined steps, the book has demonstrated in detail and through examples the methodology communities should employ to instigate, develop, and integrate a GI plan into the wider context of their community planning, such as their comprehensive plans and zoning regulations. It has shown the importance of setting goals and gathering the right data and how to create useful maps that are a vital tool to inform strategies.

The Green Infrastructure Center and Esri have coordinated to create a range of maps that can be used not just to identify where green assets might be located, but also to cross-reference themes through the use of overlays as a vital tool for community involvement and to help planners and local authorities visualize problems and their potential solutions. The mapping process also uses maps to surface opportunities, overlay connections, and provide a solid, scientific basis for moving forward to implementation.

In this context, the lead author has published a series of other guides and books showing how to merge GI conservation into comprehensive plans and community engagement, while making the case for GI planning. As users create new maps, models, and plans, they are encouraged to share them on the book's project website and at Esri user conferences and other venues.[15]

One day, it is hoped that thinking of our landscapes as our GI will be second nature. In such a worldview, no one would need to ask authorities to reconsider the route of a new highway to avoid a high-ranking core, as the issue already would have been a prime concern in the road's initial siting; no one would cut down a large forested core or fill a wetland to build subdivisions when less sensitive or redevelopment lands are available; future climate change would be automatically included in decisions made today; and degraded landscapes would be planted and reconnected for future generations.

As technology improves, we will see even finer-grained layers of information. As processing speeds increase, we will find answers faster than ever before. But data are only data. Data are helpful only if they serve a necessary purpose. Let us start by asking the right questions, first by asking, "Is your green infrastructure connected?" Ultimately, we seek a world of intact, healthy, and resilient landscapes. We hope this book has helped you better connect your world, as there is only one earth and we are the stewards for all that dwell upon it.

Enjoy the journey. Credit: Green Infrastructure Center Inc.

Notes

1. Anderson, Mark G., Melissa Clark, and Arlene Olivero Sheldon, "Estimating climate resilience for conservation across geophysical settings." *Conservation Biology* 28, no. 4 (2014): 959–970.

2. Malvern Avenue Street Scape Plan. City of Hot Springs. https://www.cityhs.net/DocumentCenter/View/7045. Website accessed February 10, 2018.

3. Hot Springs Green Infrastructure Plan. Green Infrastructure Center Inc. http://www.gicinc.org/PDFs/Hot%20Springs%20AR%20GI%20Study%20and%20Plan.pdf. Website accessed February 1, 2018.

4. European Green Belt Fact Sheet. http://www.europeangreenbelt.org/fileadmin/content/downloads/Fact-sheet_EGB_initiative_20160913.pdf. Website accessed February 28, 2017.

5. Directorate-General for Environment (European Commission). *Supporting the Implementation of Green Infrastructure: Final Report*. Publication: ENV.B.2/SER/2014/0012 (Rotterdam: European Commission, May 31, 2016). Accessed February 28, 2017.

6. Kilbane, Simon, "Green infrastructure: Planning a national green network for Australia." *Journal of Landscape Architecture* 8, no. 1 (2013): 64–73.

7. Law 5/2014, of July 25, of the Spatial Planning, Urban Planning and Landscape, of the Autonomous Region of Valencia [2014/7303].

8. Muñoz-Criado, A., and Vincente Domenech Gregori, *Comunitat Valenciana 2030: Síntesis de la Estrategia Territorial* [The Valencia Region in 2030: A Synthesis of the Regional Spatial Plan] (Valencia: Generalitat Valenciana, 2013).

9. Muñoz-Criado, Arancha, and Vicente Domenech. "Green Infrastructure Planning at Multiple Levels of Scale: Experiences from the Autonomous Region of Valencia, Spain." Scale-sensitive Governance of the Environment (2014): 283–301.

10. Strategic Regional Plan for the Region of Valencia 2030. DECREE 1/2011, of January 13, of the Regional Government of Valencia by which approves the Spatial Strategy of the Region of Valencia. [2011/235]

11. Law 4/2004, of 30 June, on Land Use Planning and Landscape Protection of the Autonomous Region of Valencia.

12. Muñoz-Criado, A., et al., "La Nueva Política de Paisaje de la Comunitat Valenciana: Programas y Actuaciones 2011–2015" [The New Valencia Landscape Policy: Programs and Interventions, 2011–2015].

13. Muñoz-Criado, A., et al., "Guía metodológica de Estudios de Paisaje" [Methodology and Guidelines for Landscape Studies]. 2012. http://datos.bne.es/edicion/a5237728.html

14. Muñoz-Criado, A., *Plan de la Huerta de Valencia. Un paisaje cultural milenario*. Vol. 1, *Estrategias de preservación y gestión*. Vol. 2, *Conclusiones del proceso de participación pública* (2009).

15. Firehock, Karen, *Strategic Green Infrastructure Planning: A Multi-Scale Approach* (Washington, DC: Island Press, 2015).

Appendix A

A brief history of landscape connectivity theory and modeling

For those wanting to understand the historical underpinnings of landscape connectivity thinking, we provide an abbreviated look at the key developments in multiple fields that led to green infrastructure network design. Understanding that there is a relationship amongst soils, geology, climate, plants, insects, microbes, and animals dates to the beginning of human history. For tens of thousands of years, aboriginal peoples made observations about their environment and the relationships between physical factors that shaped how they live and survive.

Of necessity, this review starts a bit later than the Paleolithic period, but classical Greeks reported the effects of a degraded environment, and the Roman Lucretius wrote about them in the last century BCE. Concerns about the effects of a purely urban life were evident in many writers, from Plato to St. Augustine to Thomas More's *Utopia*, published in 1516. But the science of environmental impacts on human and other species really began with the Prussian explorer and geographer Alexander von Humboldt (1769–1859). He was one of the earliest explorers to study the relationship between organisms and their environment. He observed relationships between plant species and climate, and described vegetation zones using latitude and altitude. His 1807 essay, "Geography of Plants," hypothesized insightfully that varying physical conditions affected the distribution of organic life.[1] Today, this discipline is known as geobotany. His 3D illustrations and keen observations of the natural world, down to the smallest details, are still renowned throughout the world.

Charles Darwin and Alfred Wallace revolutionized thinking about evolution when they presented papers explaining their theories of natural selection at the Linnean Society of London in August 1858. Although Darwin's notebooks from the 1830s showed earlier work on this topic, the theory gained widespread attention in 1859 when Darwin published *On the Origin of Species*. The theory of natural selection followed related ideas that prior scholars had considered (including inspiration from Thomas Robert Malthus, an early economist). But it was Darwin's evidence-based explanation of natural selection that led to wider understanding, although many peers considered it controversial. Eventually, most scholars accepted the posited nexus between environment and species' adaptation.

The importance of landscape connectivity was further popularized by early landscape architects, such as Frederick Law Olmsted (1822–1903) and his mentor and partner, the British architect and landscape designer Calvert Vaux (1824–1895). Their designs in the latter part of the 19th century marked a pivotal point in America's landscape conservation movement, as connectivity became a driving force in Olmsted's work. His interconnected park systems for cities such as Buffalo, New York, Milwaukee, Wisconsin, and Louisville, Kentucky, as well as his famous 1885 "Emerald Necklace" of connected parks for Boston, Massachusetts, showed the importance of linked networks of green spaces. In 1899, Olmsted's partner Charles Eliot proposed a visionary green space framework for the 250-square-mile Boston metropolitan area. This green network included three major rivers and six large, mostly connected, open spaces at the edges of the region, as well as a public beach.

Warren H. Manning (1860–1938) was another influential landscape architect and student of Olmsted, as well as a cofounder of the American Society of Landscape Architects. Like Olmsted, Manning was a strong proponent of saving wild landscapes for both people and wildlife, and he recognized the importance of community-based participatory design. In 1912, he used a light table to overlay maps and depict landscape relationships for his plan for Billerica, Massachusetts.[2] He then moved to a national scale, and from 1914 to 1916 he advocated for "A National Plan for the United States," in which he proposed land conservation using a series of overlay maps to show key relationships.[3] He combined both social data and geography, synthesizing 2,000 data sources to show relationships among soils, water, climate, business, land use, recreation, population, transportation, energy and minerals, and agriculture and forests. He did not publish his entire plan but did produce a "National Plan study brief" in *Landscape Architecture* magazine in July 1923.

Another early thinker in the movement to design with nature in mind was Austrian architect Richard Neutra (1892–1970). In 1954, his book *Survival Through Design* advocated consideration of site-scale thinking; understanding of the natural site conditions and constraints before, during, and after construction; and siting buildings to be integrated into the landscape. Frank Lloyd Wright (1867–1959) was another giant in this field of sustainable design. Wright designed more than 1,000 structures, perhaps the most famous of which is the 1935 Falling Waters house in Pennsylvania, in which the natural stream becomes part of the home's design. Although today we would not advocate developing over a stream, the incorporation of water was unique in American architecture of that time.

Although these ideas of blending into the landscape are not new—recall the ancient Anasazi cliff homes of Utah, Colorado, and New Mexico constructed many centuries ago—these architects and landscape architects advanced the notion that humanity's best work was not achieved by the subjugation of nature but by working with it. Meanwhile, notions of avoiding disturbance altogether and the importance of connectivity were evolving within the fields of ecology, conservation biology, and landscape architecture.

Anasazi peoples blended their homes into the landscape, as seen in this cliff dwelling from the 13th century.
Credit: Green Infrastructure Center Inc.

The American conservation movement has also played a major role in setting aside large landscapes to provide habitats for wildlife and parks to provide access for citizens. Visionaries such as John Muir, a naturalist and early advocate for preserving wilderness, led efforts to build awareness for conserving America's landscapes. He lobbied for the act to establish the National Park system in 1890, which created Yosemite National Park, among others. His contemporary, Gifford Pinchot, believed in using nature to serve humanity, but he also advocated for conservation of the nation's forests through planned use and renewal. He served as the first chief of the US Forest Service, where he put many of his ideas into practice. Over time, the national forests in the US became recognized for many additional values, including watershed protection, wildlife habitat, hunting, fishing, and general recreation.

Even as the American conservation movement grew and more lands were protected, scientists and planners began to recognize the importance of systems thinking. Early work by English botanist Arthur Tansley (1871–1955) led to the notion of ecosystems. In 1935, he published "The Use and Abuse of Vegetational Concepts and Terms," where he introduced the ecosystem concept.[4] He defined an ecosystem as a community of living organisms in conjunction with nonliving components of the environment, such as air, water, and minerals and soil, interacting as a system. He also understood that such biotic and abiotic components were linked together through nutrient cycles and energy flows.

Another key concept emerged from Alex Watt (1892–1985). He thought about plant species within bounded communities that are distributed in patches that form mosaics across a landscape. Patches are distinct areas of similar types of habitat that differ significantly from adjacent landscapes and are often dependent on delicate climate factors or a specific underlying rock type and hydrology. Watt called this theory patch dynamics in his 1947 address to the British Ecological Society.[5] He described each plant community as a "working mechanism" that constituted a dynamic space–time mosaic that "maintains and regenerates itself." His groundbreaking work led to greater understanding of the spatial dynamics of plant communities and those factors that determine specific plant patterns under specific conditions.

Watt's theory was based on observations that ecosystems are spatially heterogeneous—they contain a diverse and unevenly distributed mix of organisms and resources that offer mutual interdependence. Patch dynamics became a dominant theme in ecology beginning in the late 1970s when Levin and Paine further developed Watt's work.[6] Simon Levin, a theoretical ecologist, and Robert Paine, an empirical ecologist, created a mathematical patch-dynamics model to predict community structure. This was based on Paine's earlier field experiments on the role of predators in fostering community diversity in the rocky intertidal habitats of the US Pacific Northwest.[7]

Habitat patches are relatively homogeneous, nonlinear areas of natural land cover. Credit: Green Infrastructure Center Inc.

John Thomas Curtis (1913–1961), an American botanist and plant ecologist, further showed that plant communities are composed of species that respond individualistically to environmental and disturbance gradients. He documented dramatic changes to populations caused by the elimination of forests and native grasslands and demonstrated how species are lost because of habitat fragmentation and not just overall habitat loss.[8] As a result of all this work, ecosystems today are thought of as a mosaic of patches. Patch size, shape, duration, and boundaries are mutable and of varying sizes and sensitivities to change. Disturbances such as wind, floods, and fire can irrevocably disrupt their communities.

In 1958, studies by Carl Barton Huffaker emphasized the importance of habitat connectivity and the role of habitat corridors. He examined predator–prey relationships and the importance of the spatial structure of habitats as they affected population dynamics, especially the role of corridors in allowing species dispersal and survival.[9] Corridors provide a way for wildlife to cross the landscape and increase potential connectivity between habitat patches, which allows for greater movement and intermingling of populations. This intermingling, in turn, allows for greater genetic diversity, as well as for repopulation of previously disturbed areas. The role of corridors for providing connectivity has increasingly been recognized, as has the understanding that the corridors themselves provide habitat.

Early planning involving such natural corridors is seen in the work of landscape architect Phil Lewis, who in 1964 proposed a statewide network of green spaces and environmental corridors along rivers and wetlands in Wisconsin. It is notable that his map included both natural and cultural resources.

Robert McArthur and Edward O. Wilson helped lay the foundational principles for understanding the impacts of habitat fragmentation and species isolation in their 1967 book *The Theory of Island Biogeography*, in which an "island" is any area of habitat suitable for a specific ecosystem, surrounded by an expanse of unsuitable habitat. They found these ecological islands affect the species' richness. Wilson tested his theory of island biogeography in the Florida Keys, where they found that species within disturbed mangrove areas recovered faster, the more proximate they were to similar habitats.[10]

Regulations such as the Endangered Species Act (ESA) of 1969 and its 1973 update advanced the American conservation movement and drew attention to the need for sustained conservation and protection for at-risk species in order to help them recover.[11] The ESA requires planning for habitat conservation by supporting the protection of habitat for species identified as "endangered."

The idea of a *metapopulation* is another central concept, coined by Richard Levins in 1969 to describe a group of spatially separated populations of the same species that interact at some level. This concept is applied in conservation biology to the suitability of landscape patches as habitat and the probability for extinction of a species in a patch or its recolonization.

For greater detail on the evolution of hierarchical patch dynamics, see *From Balance of Nature to Hierarchical Patch Dynamics: A Paradigm Shift in Ecology*, in which Jianguo Wu and Orie L. Loucks synthesize the shift from thinking of nature as an ordered and harmonious system to understanding it in terms of a hierarchical patch-dynamics paradigm in which ecological processes operate over a wide range of spatial, temporal, and organizational scales, and at varying rates of change.[12]

R. T. T. Forman is a biologist and botanist whose work in plant, avian, and forest ecology has tested the effect of patch size on biodiversity, both in his 1962 work on the hierarchical distributions of plant species and in his 1976 study of old-growth woods within farmland. These studies form the basis for much of his later work, which includes a focus on the ecology of roads, as well as on the spatial configurations of organisms, structures, and the built environment. His 1995 book, *Land Mosaics: The Ecology of Landscapes and Regions*, illustrates habitat patterns and landscape connectivity and how these patterns contribute to species' movement.[13]

The fields of planning and landscape architecture are also central to GI. Ian McHarg (1920–2001) is best known for his 1969 book *Design with Nature*, in which he laid out principles for understanding the landscape as an interrelated system. His notion of designing with nature drew on early innovators, such as Scottish biologist Patrick Geddes. Geddes trained with Thomas Huxley, who was also Charles Darwin's collaborator. Geddes sought to understand humans' interaction with the natural environment and how that shaped settlement patterns, the economy, and ecology. Rather than reshaping the landscape solely to suit humans' needs, McHarg proposed that working with existing landforms and drainage patterns led to more harmonious developments. He first overlaid existing soils, slopes, drainages, and vistas to understand the functions of the natural landscape first and thus intentionally avoid damaging or disrupting those natural systems.[14]

Another key development for landscape ecology and planning was the gap analysis method recommended by J. Michael Scott and others in 1993, which created a systematic analysis for habitat distribution and biodiversity hotspots.[15] Gap analysis responded to growing awareness that even large parks, such as Yellowstone National Park in the US, could not fully contain and conserve all species. Gap analysis requires the demarcation of habitat types and a comparison of those distributions with lands under protective status on a state-by-state basis to uncover "gaps" in protection—in other words, those habitat types that were underrepresented in each state's system of parks and conserved lands.

Greenway design is yet another discipline in the field of planning related to corridors and connectivity that includes a human-centric focus, mostly for outdoor recreation. The term greenway arose from combining of the words *greenbelt* and *parkway*. Greenways function as linear parks and were possibly mentioned first in a William Whyte monograph, "Securing Open Space for Urban America," in 1959.[16]

In 1995, Charles Little further popularized the greenways movement in his book *Greenways for America*, which focused on connecting trails along river corridors through such greenways. These were loosely defined as "linear open spaces that preserve and restore nature in cities, suburbs and rural areas . . . to link parks and open spaces and provide corridors for wildlife migration."[17]

In 2006, Paul Cawood Hellmund and Daniel Smith updated the notion of greenways to incorporate the connection of large-scale habitats in their book *Designing Greenways: Sustainable Landscapes for Nature and People*. They offered a more useful and broader description of GI that built on the greenways movement by adding principles of landscape ecology from Forman's earlier work.[18]

A key element for all this work is the need for interdisciplinary focus. Noted landscape architect Carl Steinitz coined the term *geodesign* in 2012 to characterize his longstanding process for integrating various models for landscape assessment and intervention. According to Steinitz, geodesign is a framework and process for multidisciplinary planning and practice rather than a particular discipline.[19] Geodesign focuses on the collaborative nature of planning, recommending the inclusion of different perspectives and types of knowledge when designing a landscape, whether the aims are protection, restoration, or reconnection.[20]

Geodesign also suggests that there are different considerations for what is important depending on the scale of the focus. For example, knowing the exact types of salamanders in an area might be important for a site-scale plan, whereas knowing generally which areas likely include characteristics that are suitable for salamander habitat might be more important when crafting a habitat model for a large area, such as an entire county or watershed.

This brief overview and references have provided a foundation for those who want to learn more about the scientific underpinnings of green infrastructure network design. Readers are encouraged to investigate the references provided in the Notes section for a fuller understanding of the rich scientific inquiry that has brought us to our practice of landscape ecology and ecological network design.

Notes

1. Von Humboldt, Alexander, and Aimé Bonpland, *Essay on the Geography of Plants* (Chicago: University of Chicago Press, 2010).
2. Steinitz, Carl, *A Framework for Geodesign: Changing Geography by Design* (Redlands, CA: Esri Press, 2012).
3. Ibid.
4. Tansley, Arthur G., "The use and abuse of vegetational concepts and terms." *Ecology* 16, no. 3 (1935): 284–307.
5. Watt, Alex S., "Pattern and process in the plant community." *Journal of Ecology* 35, no. 1/2 (1947): 1–22.

6. Levin, Simon A., and Robert T. Paine, "Disturbance, patch formation, and community structure." *Proceedings of the National Academy of Sciences* 71, no. 7 (1974): 2744–2747.

7. Ibid.

8. Curtis, John T., "The modification of mid-latitude grasslands and forests by man." In *Man's Role in Changing the Face of the Earth* (Chicago: University of Chicago Press, 1956), 721–736.

9. Huffaker, C. B., "Experimental studies on predation: Dispersion factors and predator–prey oscillations," *Hilgardia* 27, no. 14 (1958): 343–383.

10. MacArthur, Robert H., and Edward O. Wilson, *The Theory of Island Biogeography* (Princeton, NJ: Princeton University Press, 2001).

11. Noss, Reed F., Michael O'Connell, and Dennis D. Murphy, *The Science of Conservation Planning: Habitat Conservation under the Endangered Species Act* (Washington, DC: Island Press, 1997).

12. Wu, Jianguo, and Orie L. Loucks, "From balance of nature to hierarchical patch dynamics: A paradigm shift in ecology." *Quarterly Review of Biology* 70, no. 4 (1995): 439–466.

13. Forman, R. T. T., *Land Mosaics: The Ecology of Landscapes and Regions* (Cambridge: Cambridge University Press, 1995).

14. McHarg, Ian L., and Lewis Mumford, *Design with Nature* (Garden City, NY: The Natural History Press for the American Museum of Natural History, 1969).

15. Scott, J. Michael, Frank Davis, Blair Csuti, Reed Noss, Bart Butterfield, Craig Groves, Hal Anderson, et al., "Gap analysis: A geographic approach to protection of biological diversity." *Wildlife Monographs* (1993): 3–41.

16. Whyte, William Hollingsworth, "Securing open space for urban America: Conservation easements." *Urban Land Institute* 2, no. 3 (1959).

17. Little, Charles E., *Greenways for America* (Baltimore: Johns Hopkins University Press, 1995).

18. Hellmund, Paul Cawood, and Daniel Smith, *Designing Greenways: Sustainable Landscapes for Nature and People* (Washington, DC: Island Press, 2013).

19. Steinitz, *A Framework for Geodesign*.

20. Ibid.

Appendix B

Technical appendix: Core attributes

What is a core and how is it made? Cores are intact habitat areas at least 100 acres in size and at least 200 meters wide. They are derived from the 2011 National Land Cover Database. Potential cores areas are selected from land cover categories not containing the word *developed* or those categories associated with agriculture uses (crop, hay, and pasture lands). The resulting areas are tested for size and width and converted into unique polygons. These polygons are then overlaid with a diverse assortment of physiographic, biological, and hydrographic layers to use in computing a core quality index. The resulting attributes of these polygons are described below.

Internal reserved field names used in GIS to describe attributes

The following four fields are created automatically when using ArcGIS software:
> **OBJECTID** *(OBJECTID)*—an internal ArcGIS ID value for each polygon
> **Shape** *(Shape)*—an internal ArcGIS description ("polygon") for the feature type in the intact core file
> **Shape_Area** *(Shape_Area)*—the area in meters of a core
> **Shape_Length** *(Shape_Length)*—the perimeter in meters of a core

The following value field is to identify each unique core:
> **Value** *(Unique Core Identifier)*—a unique identifier for each core. It is used to relate core features to the Supplemental Attribute Table.

Biology

> **BiodiversityPriorityIndex** *(Biodiversity Priority Index)*—the intact core areas were overlaid with the Priority Index Layer (10 km) resolution surface described in C. N. Jenkins, K. S. van Houtan, S. L. Pimm, & J. O. Sexton, "US protected lands mismatch biodiversity priorities," *Proceedings of the National Academy of Sciences* 112, no. 16

(2015): 5081–5086, http://www.pnas.org/cgi/doi/10.1073/pnas.1418034112. The Priority Index score is a summary for each of 1,200 endemic species of the proportion of species range that is unprotected divided by the area of the species' range. Values are summed across all endemic species within a taxonomic group and across all taxonomic groups. Cores falling within a priority index category are assigned that priority index value. Note that the nominal resolution of the Priority Index data is 10 km. Cores may or may not have endemic species or collections of endemic species within them.

EcolSystem_Redundancy *(Ecological System Redundancy)*—measures the number of TNC Ecoregions Systems in which a GAP Level 3 Ecological Systems occurs. The higher the number, the more Ecoregions an Ecological System appears in and the greater its redundancy. Cores are scored with lowest redundancy value appearing within them. Low and very low redundancy values represent cores containing unique Ecological Systems. This analysis reproduces the work by Jocelyn L. Aycrigg, Anne Davidson, Leona K. Svancara, Kevin J. Gergely, Alexa McKerrow, and J. Michael Scott, "Representation of ecological systems within the protected areas network of the continental United States," *PLOS ONE* 8, no. 1 (2013): e54689, applied to finer-resolution TNC Ecoregions units.

EndemicSpeciesMax *(Endemic Species Max)*—the maximum count of endemic species (trees, freshwater fish, amphibians, reptiles, birds, mammals) per core when overlaid with an Endemic Species dataset (10 km) resolution from https://biodiversityMapping.org.

TNC_Ecoregion_Maj *(TNC Ecoregion Name)*—the name of the predominant TNC ecoregion present within a core. From The Nature Conservancy's (TNC) Terrestrial Ecoregions database. There are 68 TNC ecoregions covering the lower 48 US states, http://maps.tnc.org/gis_data.html#TerrEcos.

Evaluation

Score *(Core Score)*—the core quality index value based on geometric values and soil variety, endemic species max, biodiversity priority index, and ecological systems redundancy. This calculation is based on the Green Infrastructure Center's (http://gicinc.org) scoring methodology in its Practitioner's Guides.

Score_Bio *(Score Using High Biologic Component Weight)*—the core quality index value based on geometric values and soil variety, endemic species max, biodiversity priority index, and ecological systems redundancy. This approach reduces the importance of the geometric value area from .4 to .2 and increases the importance of biological elements by .2 overall. This is a modification of the Green Infrastructure Center's (http://gicinc.org) scoring methodology in its Practitioner's Guides.

Score_Bio_byTNCEcoregion *(Score Using High Biologic Component Weight by TNC Ecoregion)*—the core quality index value based on geometric values and soil variety, endemic species max, biodiversity priority index, and ecological systems redundancy, performed within individual TNC Ecoregion units. This approach reduces the importance of the geometric value area from .4 to .2 and increases the importance of biological elements by .2 overall. This is a modification of the Green Infrastructure Center's (http://gicinc.org) scoring methodology in its Practitioner's Guides.

Score_byTNCEcoregion *(Score Calculated by TNC Ecoregion)*—the core quality index value based on geometric values and soil variety, endemic species max, biodiversity priority index, and ecological systems redundancy, performed within individual TNC Ecoregion units. This calculation is based on the Green Infrastructure Center's (http://gicinc.org) scoring methodology in its Practitioner's Guides.

Score_NHDPlus *(Score with NHDPlus Flow Greater Than 1 cfs)*—the core quality index value based on geometric values and soil variety, endemic species max, biodiversity priority index and ecological systems redundancy for cores containing an NHDPlus FTYPE (StreamRiver) and Q0001A => 1.0 in cubic feet per second. This represents a subset of all cores. This calculation is based on the Green Infrastructure Center's (http://gicinc.org) scoring methodology in its Practitioner's Guides.

Score_NHDPlus_Bio *(Score Using High Biologic Component Weight and NHDPlus Flow)*—the core quality index value based on geometric values and soil variety, endemic species max, biodiversity priority index, and ecological systems redundancy for cores containing an NHDPlus FTYPE (StreamRiver) and Q0001A => 1.0 in cubic feet per second. This approach reduces the importance of the geometric value area from .4 to .2, and increases the importance of biological elements by .2 overall. This represents a subset of all cores. This is a modification of the Green Infrastructure Center's (http://gicinc.org) scoring methodology in its Practitioner's Guides.

Score_NHDPlus_Bio_byTNCEcoregion *(Score Using High Biologic Component Weight and NHDPlus Flow by TNC Ecoregion)*—the core quality index value based on geometric values and soil variety, endemic species max, biodiversity priority index, and ecological systems redundancy for cores containing an NHDPlus FTYPE (StreamRiver) and Q0001A => 1.0 in cubic feet per second, performed within individual TNC Ecoregion units. This approach reduces the importance of the geometric value area from .4 to .2 and increases the importance of biological elements by .2 overall. This represents a subset of all cores. This is a modification of the Green Infrastructure Center's (http://gicinc.org) scoring methodology in its Practitioner's Guides.

Score_NHDPlus_byTNCEcoRegion *(Score with NHDPlus Flow Greater Than 1 cfs by TNCEcoRegion)*—the core quality index value based on geometric values and soil variety, endemic species max, biodiversity priority index, and ecological systems redundancy for cores containing an NHDPlus FTYPE (StreamRiver) and Q0001A => 1.0 in cubic feet per second, performed within individual TNC Ecoregion units. This represents a subset of all cores. This calculation is based on the Green Infrastructure Center's (http://gicinc.org) scoring methodology in its Practitioner's Guides.

Score_User *(Score Generated by User Interaction)*

Geometry

Acres *(Acres)*—core area in acres.

Class *(Core Size Class)*—the size class for each core (area − water). If < 100 acres = fragment, if < 1,000 = small, if < 10,000 = medium, if > 10 K = large.

Compactness *(Compactness)*—the ratio between the area of the core and the area of a circle with the same perimeter as the core.

PA_Ratio *(Perimeter/Area Ratio)*—the perimeter/area ratio of core.

Thickness *(Thickness)*—represents the deepest or thickest point within each core. Essentially, it is the radius (in cells) of the largest circle that can be drawn within each core without including any cells outside the core. Cores with greater "depth or thickness" are preferred because they represent larger and potentially safer interior core areas.

Land cover

HM_Mean *(Human Modified Index Mean Value)*—the mean of the Theobald Human Modified values appearing in a core. A measure of the degree of human modification, the index ranges from 0.0 for a virgin landscape condition to 1.0 for the most heavily modified areas. The average value for the US is 0.375. The data used to produce these values should be both more current and more detailed than the NLCD used for generating the cores. Emphasis was given to attempting to map, in particular, energy-related development. D. M. Theobald, "A general model to quantify ecological integrity for landscape assessment and US Application," *Landscape Ecology* 28 (2013): 1859–1874, doi: 10.1007/s10980-013-9941-6.

HM_Std *(Human Modified Index Standard Deviation)*—the standard deviation of the Theobald Human Modified values appearing within a core. A measure of the degree of human modification, the index ranges from 0.0 for a virgin landscape condition to 1.0 for the most heavily modified areas. The average value for the US is 0.375. The data used to produce these values should be both more current and more detailed than the NLCD used for generating the cores. Emphasis was given to attempting to map, in particular, energy-related development. D. M. Theobald, "A general model to quantify ecological integrity for landscape assessment and US application," *Landscape Ecology* 28 (2013): 1859–1874, doi: 10.1007/s10980-013-9941-6.

NLCD_Forested_Pct *(Percentage NLCD Forested)*—the percentage of Forested land (NLCD deciduous [41], evergreen [42], or mixed [43]) within a core, expressed as a number between 0 and 100. These are from the 2011 National Land Cover Database, **https://www.mrlc.gov/data?f%5B0%5D=year%3A2011**.

NLCD_GrasslandHerb_Pct *(Percentage NLCD Grassland/Herbaceous)*—the percentage of Grass/Herbaceous land (NLCD grass/herb [71], sedge/herb [72], lichens

[73], or moss (74)) within a core, expressed as a number between 0 and 100. These are from the 2011 National Land Cover Database, https://www.mrlc.gov/data?f%5B0%5D=year%3A2011.

NLCD_ShrubScrub_Pct *(NLCD Shrub and Scrub)*—the percentage of Shrub/Scrub land (NLCD dwarf scrub [51] or shrub/scrub [52]) within a core, expressed as a number between 0 and 100. These are from the 2011 National Land Cover Database, https://www.mrlc.gov/data?f%5B0%5D=year%3A2011.

NLCD_WetlandsHerbWet_Pct *(Percentage NLCD Wetlands/Herbaceous)*—the percentage of Wetlands (NLCD woody wetlands [90] or emergent herbaceous [95]) within a core, expressed as a number between 0 and 100. These are from the 2011 National Land Cover Database, https://www.mrlc.gov/data?f%5B0%5D=year%3A2011.

Morphology

ELU_BIO_De *(ELU Bioclimate Description)*—the name of the primary Ecological Land Unit bioclimate type within each core. An Ecological Land Unit is an area of distinct bioclimate, landform, lithology, and land cover. The data are available from the USGS at http://rmgsc.cr.usgs.gov/outgoing/ecosystems/Global.

ELU_GLC_De *(ELU Global Landcover Description)*—the name of the primary Ecological Land Unit land cover type within each core. An Ecological Land Unit is an area of distinct bioclimate, landform, lithology, and land cover. The data are available from the USGS at http://rmgsc.cr.usgs.gov/outgoing/ecosystems/Global.

ELU_ID_Maj *(ELU Majority)*—the primary Ecological Land Unit appearing within a core. An Ecological Land Unit is an area of distinct bioclimate, landform, lithology, and land cover. The data are available from the USGS at http://rmgsc.cr.usgs.gov/outgoing/ecosystems/Global.

ELU_LIT_De *(ELU Lithology Description)*—the name of the primary Ecological Land Unit lithology type within each core. An Ecological Land Unit is an area of distinct bioclimate, landform, lithology, and land cover. The data are available from the USGS at http://rmgsc.cr.usgs.gov/outgoing/ecosystems/Global.

ELU_LSF_De *(ELU Landform Description)*—the name of the primary Ecological Land Unit landform type within each core. An Ecological Land Unit is an area of distinct bioclimate, landform, lithology, and land cover. The data are available from the USGS at http://rmgsc.cr.usgs.gov/outgoing/ecosystems/Global.

ELU_SWI *(ELU Shannon Weaver Diversity Index)*—the Shannon-Weaver diversity index of the Ecological Land Units appearing within a core. An Ecological Land Unit is an area of distinct bioclimate, landform, lithology, and land cover. The data are available from the USGS at http://rmgsc.cr.usgs.gov/outgoing/ecosystems/Global. Greater diversity is frequently associated with better habitat potential.

ERL_Descriptor *(ERL Description)*—the name of the primary Theobald Ecologically Relevant Landform within a core. From D. M. Theobald, D. Harrison-Atlas, W. B. Monahan, and C. M. Albano, "Ecologically-relevant landforms and physiographic diversity for climate adaptation planning," *PLOS ONE* 10, no. 12 (2015):e0143619, doi: 10.1371/journal.pone.0143619.

ERL_Maj *(ERL Majority Type)*—the dominant landform by area appearing with in a core from Theobald's Ecologically Relevant Landforms. From D. M. Theobald, D. Harrison-Atlas, W. B. Monahan, and C. M. Albano, "Ecologically-relevant landforms and physiographic diversity for climate adaptation planning," *PLOS ONE* 10, no. 12 (2015):e0143619, doi: 10.1371/journal.pone.0143619.

ERL_SWI *(ERL Shannon-Weaver Diversity Index)*—the Shannon-Weaver diversity index of the Theobald Ecologically Relevant Landforms appearing within a core. From D. M. Theobald, D. Harrison-Atlas, W. B. Monahan, and C. M. Albano, "Ecologically-relevant landforms and physiographic diversity for climate adaptation planning," *PLOS ONE* 10, no. 12 (2015):e0143619, doi: 10.1371/journal.pone.0143619. Greater diversity is frequently associated with better habitat potential.

GAP_EcolSystem_L3_Maj *(GAP Ecological System Level 3 Majority)*—the primary GAP Level 3 Ecological System appearing within a core. The USGS GAP Level 3 code references the Ecological System classification element developed by NatureServe, which is mainly focused on habitat identification. Roughly 540 of the 590 ecological systems in the GAP database appear in these data. See **http://gapanalysis.usgs.gov/gaplandcover/data/land-cover-metadata/#5** for more information.

GAP_EcolSystems_L3_SWI *(Ecological System Shannon-Weaver Diversity Index)*—the Shannon-Weaver diversity index of GAP Level 3 Ecological Systems within a core. The USGS GAP Level 3 code references the Ecological System classification element developed by NatureServe, which is mainly focused on habitat identification. Greater diversity is frequently associated with better habitat potential. Roughly 540 of the 590 ecological systems in the GAP database appear in these data. See **http://gapanalysis.usgs.gov/gaplandcover/data/land-cover-metadata/#5** for more information.

Landform_Maj *(Landform Description [Esri])*—the primary local landform name within a core from Karagulle/Frye method. These are "local" representations of Hammond's Landform Classification categories.

Soil_SWI *(Soil Shannon-Weaver Diversity Index)*—the Shannon-Weaver diversity index of SSURGO, MUKEY values appearing within a core. These are map units from the Dept. of Agriculture's National Cooperative Soil Survey. Diversity in soils is a surrogate for diversity in habitat potential. Greater diversity is frequently associated with better habitat potential. Data have been collected over the past 100 years and are most intensively mapped in areas with high agricultural potential. Data are missing for many national forests, national parks, and arid lands.

Soil_Variety *(Soil Mukey Variety)*—the number of different SSURGO MUKEY units appearing within a core. These are map units from the Dept. of Agriculture's National Cooperative Soil Survey. Variety in soils is a surrogate for diversity in habitat potential. Greater variety should equate to greater habitat potential. Data have been collected over the past 100 years and are most intensively mapped in areas with high agricultural potential. Data are missing for many national forests, national parks, and arid lands.

Topo_Std *(Elevation Variability Standard Deviation)*—the standard deviation of the topographic diversity from NED 30-meter resolution, using zonal statistics within a core. The presumption is that the larger the deviation, the better it is for habitat potential.

Water related

NHDPlusFlowLenFtPerAcre *(NHDPlus Flow Length [ft] per Core Acre)*—the length of NHDPlus FTYPE (StreamRiver) and Q0001A => 1.0, in cubic feet within a core/core area in acres. This is a measure of features with running water as modeled in the NHDPlusV2 database from the EPA and USGS, https://www.epa.gov/waterdata/nhdplus-national-hydrography-dataset-plus. This variable is to distinguish hydrologic features with active flows from intermittent, artificial, or pipeline or canal features.

NHDPlusFlowLen_ft *(NHDPlus Stream and River Length [ft] Flow Greater Than 1 cfs)*—the length of NHDPlus FTYPE (StreamRiver) and Q0001A => 1.0 in cubic feet within each core. This is a measure of features with running water as modeled in the NHDPlusV2 database from the EPA and USGS, https://www.epa.gov/waterdata/nhdplus-national-hydrography-dataset-plus. This variable is to distinguish hydrologic features with active flows from intermittent, artificial, or pipeline or canal features.

Strm_Len_ft *(Stream Length All [ft] [NHD])*—stream length (all types) in feet within a core. This captures the broadest possible collection of hydrologic features from the National Hydrography Dataset. These may overrepresent the presence and availability of water, particularly in the Southwest.

Strm_LenPerAcre *(Stream Length All [ft] per Core Acre [NHD])*—stream length (all types) in feet within a core/core area in acres. This captures the broadest possible collection of hydrologic features from the National Hydrography Dataset. These may over represent the presence and availability of water, particularly in the Southwest.

WaterBodiesPct *(Percentage Water Area per Core [NWI])*—the percentage of water features (from NHD—includes lakes, ponds, reservoirs, SeaOcean, StreamRiver, and Canal/Ditch. Ftypes: 390, 436, 336, 445, and 460) within a core. This is expressed as a number ranging from **0.0** to **100**. From the USGS National Hydrographic Dataset http://nhd.usgs.gov/data.html. Cores with greater amounts of water have better habitat potential than those with less.

Water_AreaPerAcre *(Water Body Area per Core Acre [NHD])*—the percentage of water features (from NHD—includes lakes, ponds, reservoirs, SeaOcean, StreamRiver, and Canal/Ditch. Ftypes: 390, 436, 336, 445, and 460) within a core. This expressed as a number ranging from **0.0** to **1.0**. From the USGS National Hydrographic Dataset, http://nhd.usgs.gov/data.html. Cores with greater amounts of water have better habitat potential than those with less.

Wet_AreaPerAcre *(Wetlands Area per Core Acre [NWI])*—the percentage of wetlands (from NWI—includes "Estuarine and marine," "freshwater emergent," or "freshwater forested/shrub") within a core. This expressed as a number ranging from **0.0** to **1.0**. From the US Fish and Wildlife Service's National Wetlands Inventory, http://www.fws.gov/wetlands/. Cores with more wetlands have better habitat potential than those with less.

WetlandsPct *(Percentage Wetlands Area per Core [NWI])*—the percentage of wetlands (from NWI—includes "Estuarine and marine," "freshwater emergent," or "freshwater forested/shrub") within a core. This expressed as a number ranging from **0.0** to **100**. From the US Fish and Wildlife Service's National Wetlands Inventory, http://www.fws.gov/wetlands/. Cores with more wetlands have better habitat potential than those with less.

Data sources

Ecological Land Units USGS/ESRI (received 3/2016) 250-m resolution
http://rmgsc.cr.usgs.gov/outgoing/ecosystems/Global/
Local Conservation Cooperative Boundaries (downloaded 3/2016)
https://www.sciencebase.gov/catalog/item/55b943ade4b09a3b01b65d78
TNC Ecoregions (downloaded 4/2016)
http://maps.tnc.org/gis_data.html#TNClands — Ecoregional Portfolio section, US Data
Theobald's Human Modified data (received 4/2016) 90-m resolution. Referenced in
http://www.montana.edu/lccvp/documents/theobald2013.pdf
Theobald's Ecologically Relevant Landforms (downloaded 4/2016) 30-m resolution. Referenced in
http://journals.plos.org/plosone/article?id=10.1371/journal.pone.0143619
GAP Level 3 Ecological System Boundaries (downloaded 4/2016)
http://gapanalysis.usgs.gov/gaplandcover/data/download/
Local Landforms (produced 3/2016) by Deniz Karagulle and Charlie Frye 30-m* resolution. "Improved Hammond's Landform Classification and Method for Global 250-m Elevation Data," by Deniz Karagulle, Charlie Frye, Roger Sayre, Sean Breyer, Peter Aniello, Randy Vaughan, and Dawn Wright, has been successfully submitted online and is currently being given full consideration for publication in Transactions in GIS.

*We scaled the neighborhood windows from the 250-m method described in the paper, and then applied that to 30-m data in the US.

NWI—National Wetlands Inventory. Classification of Wetlands and Deepwater Habitats of the United States. FWS/OBS-79/31 (Washington, DC: US Department of the Interior, Fish and Wildlife Service, Division of Habitat and Resource Conservation, prepared October 2015).

US Fish and Wildlife Service. National Wetlands Inventory website (Washington, DC: US Department of the Interior, Fish and Wildlife Service, October 2015). http://www.fws.gov/wetlands/.

NHD—National Hydrologic Dataset, http://nhd.usgs.gov/data.html. Coordinated effort among the US Department of Agriculture, Natural Resources Conservation Service (USDA-NRCS), the US Geological Survey (USGS), and the Environmental Protection Agency (EPA). The Watershed Boundary Dataset (WBD) was created from a variety of sources from each state and aggregated into a standard national layer for use in strategic planning and accountability.

NLCD 2011—National Land Cover Database 2011, https://www.mrlc.gov/data?f%5B0%5D=year%3A2011.

Preferred Citation: NLCD 2011 citation: Homer, C.G., Dewitz, J.A., Yang, L., Jin, S., Danielson, P., Xian, G., Coulston, J., Herold, N.D., Wickham, J.D., and Megown, K., 2015, Completion of the 2011 National Land Cover Database for the conterminous United States—Representing a decade of land cover change information. *Photogrammetric Engineering and Remote Sensing*, v. 81, no. 5, p. 345–354.

NHDPlusV2—https://www.epa.gov/waterdata/nhdplus-national-hydrography-dataset-plus

Received from Charlie Frye, Esri, 3/2016. Produced by the EPA with support from the USGS.

SSURGO—Soil Survey Staff, Natural Resources Conservation Service, US Department of Agriculture. Web Soil Survey. Available online at http://websoilsurvey.nrcs.usda.gov/. Accessed 3/2016.

gSSURGO—Soil Survey Staff, Natural Resources Conservation Service, United States Department of Agriculture. Web Soil Survey. Available online at http://websoilsurvey.nrcs.usda.gov/. Accessed 3/2016, 30-m resolution

TIGER—(Roads National Geodatabase) 2015 (downloaded 1/2016).
https://www.census.gov/geo/maps-data/data/tiger-geodatabases.html
ftp://ftp2.census.gov/geo/tiger/TGRGDB15/tlgdb_2015_a_us_roads.gdb.zip

TIGER—(Rails National Geodatabase) 2015 (downloaded 1/2016).
https://www.census.gov/geo/maps-data/data/tiger-geodatabases.html
ftp://ftp2.census.gov/geo/tiger/TGRGDB15/TLGDB_2015_a_us_RAILS.gdb.zip

NOAA CCAP Coastal Change Analysis Program Regional Land Cover and Change—Downloaded by state (1/2016).
https://coast.noaa.gov/ccapftp/#/
Description: https://coast.noaa.gov/dataregistry/search/collection/info/ccapregional
30-m resolution, 2010 edition of data.

Scoring values

Default weights

- 0.4, # Acres
- 0.1, # THICKNESS
- 0.05, # TOPOGRAPHIC DIVERSITY (Standard Deviation)
- 0.1, # Biodiversity Priority Index (SPECIES RICHNESS in GIC original version)
- 0.05, # PERCENTAGE WETLAND COVER
- 0.03, # Ecological Land Unit – Shannon-Weaver Index (SOIL VARIETY in GIC original version)
- 0.02, # COMPACTNESS RATIO (AREA RELATIVE TO THE AREA OF A CIRCLE WITH THE SAME PERIMETER LENGTH)
- 0.1, # STREAM DENSITY (LINEAR FEET/ACRE)
- 0.05, # Ecological System Redundancy (RARE/THREATENED/ENDANGERED SPECIES ABUNDANCE (Number of occurrences) in GIC original version)
- 0.1, # Endemic Species Max (RARE/THREATENED/ENDANGERED SPECIES DIVERSITY (Number of unique species in a core) in GIC original version)

Bio-weights

- 0.2, # Acres=
- 0.1, # THICKNESS
- 0.05, # TOPOGRAPHIC DIVERSITY (Standard Deviation)
- 0.25, # Biodiversity Priority Index (SPECIES RICHNESS in GIC original version)
- 0.05, # PERCENTAGE WETLAND COVER
- 0.03, # Ecological Land Unit – Shannon-Weaver Index (SOIL VARIETY in GIC original version)
- 0.02, # COMPACTNESS RATIO (AREA RELATIVE TO THE AREA OF A CIRCLE WITH THE SAME PERIMETER LENGTH)
- 0.1, # STREAM DENSITY (LINEAR FEET/ACRE)
- 0.1, # Ecological System Redundancy (RARE/THREATENED/ENDANGERED SPECIES ABUNDANCE (Number of occurrences) in GIC original version)
- 0.1, # Endemic Species Max (RARE/THREATENED/ENDANGERED SPECIES DIVERSITY (Number of unique species in a core) in GIC original version)

Appendix C

Esri® Story Maps and other resources

Chapter 1
- Green Infrastructure in Action Crowdsource Story Map
- Building a National Intact Habitat Core Database Story Map
- Green Infrastructure Layers Cascade Story Map

Chapter 3
- 6-Step Guide to GI Story Map

Chapter 4
- Collection of map package layers used in the model for state download collection

Chapter 5
- Prioritize Cores to Build a Network: How to use the cores model to tease out cores that meet local goals Story Map

Chapter 6
- Assessing Risk: How to assess the cores model to tease out cores that may be at risk Story Map

Chapter 8
- Green Infrastructure in the City: Mapping connectivity at the urban scale in Charlottesville, VA

Bibliography

Alignier, Audrey, and Marc Deconchat. "Patterns of forest vegetation responses to edge effect as revealed by a continuous approach." *Annals of Forest Science* 70, no. 6 (2013): 601–609.

Anderson, Greg S., and Brent J. Danielson. "The effects of landscape composition and physiognomy on metapopulation size: The role of corridors." *Landscape Ecology* 12, no. 5 (1997): 261–271.

Anderson, Mark G., Melissa Clark, and Arlene Olivero Sheldon. "Estimating climate resilience for conservation across geophysical settings." *Conservation Biology* 28, no. 4 (2014): 959–970.

Averill, Kristine M., David A. Mortensen, Erica A. H. Smithwick, and Eric Post. "Deer feeding selectivity for invasive plants." *Biological Invasions* 18, no. 5 (2016): 1247–1263.

Aycrigg, Jocelyn L., Anne Davidson, Leona K. Svancara, Kevin J. Gergely, Alexa McKerrow, and J. Michael Scott. "Representation of ecological systems within the protected areas network of the continental United States." *PLOS ONE* 8, no. 1 (2013): e54689.

Bailey, Robert G., comp. *Description of the Ecoregions of the United States*. 2nd ed. Misc. Publ. No. 1391. Washington, DC: US Department of Agriculture, 1995.

Bailey, Robert G. *Ecoregions Map of North America: Explanatory Note*. No. 1548. Washington, DC: US Dept. of Agriculture, Forest Service, 1998.

Barnosky, Anthony D., Elizabeth A. Hadly, Jordi Bascompte, Eric L. Berlow, James H. Brown, Mikael Fortelius, Wayne M. Getz, et al. "Approaching a state shift in Earth's biosphere." *Nature* 486, no. 7401 (2012): 52–58.

Batisse, Michel. "The biosphere reserve: A tool for environmental conservation and management." *Environmental Conservation* 9, no. 2 (1982): 101–111.

Beier, Paul. "Conceptualizing and designing corridors for climate change." *Ecological Restoration* 30, no. 4 (2012): 312–319.

Beier, Paul, and Reed F. Noss. "Do habitat corridors provide connectivity?" *Conservation Biology* 12, no. 6 (1998): 1241–1252.

Beier, Paul, Wayne Spencer, Robert F. Baldwin, and Brad McRae. "Toward best practices for developing regional connectivity maps." *Conservation Biology* 25, no. 5 (2011): 879–892.

Belote, R. Travis, Matthew S. Dietz, Brad H. McRae, David M. Theobald, Meredith L. McClure, G. Hugh Irwin, Peter S. McKinley, Josh A. Gage, and Gregory H. Aplet.

"Identifying corridors among large protected areas in the United States." *PLOS ONE* 11, no. 4 (2016): e0154223.

Benedict, Mark A., and Edward T. McMahon. *Green Infrastructure: Linking Landscapes and Communities.* Washington, DC: Island Press 2006.

Bennett, Andrew F. *Linkages in the Landscape: The Role of Corridors and Connectivity in Wildlife Conservation.* Geneva, Switzerland: IUCN, 1999.

Bentz, Barbara J., Jacques Régnière, Christopher J. Fettig, E. Matthew Hansen, Jane L. Hayes, Jeffrey A. Hicke, Rick G. Kelsey, Jose F. Negrón, and Steven J. Seybold. "Climate change and bark beetles of the Western United States and Canada: Direct and indirect effects." *BioScience* 60, no. 8 (2010): 602–613.

Brock, Brent L., and Eric C. Atkinson. "Selecting species as targets for conservation planning." In *Conservation Planning*. Redlands, CA: Esri Press, 2013.

Brown, J. W. *Eco-logical: An Ecosystem Approach to Developing Infrastructure Projects.* Cambridge, MA: Research and Innovative Technology Administration, 2006.

Brown, Mark. T., J. Schaefer, and K. Brandt. *Buffer Zones for Water, Wetlands, and Wildlife in the East Central Florida Region.* Gainesville: Center for Wetlands, University of Florida, 1990.

Bushman, Ellen S., and Glenn D. Therres. *Habitat Management Guidelines for Forest Interior Breeding Birds of Coastal Maryland.* Annapolis: Maryland Dept. of Natural Resources, Forest, Park and Wildlife Services, 1988.

Cart, Julie. "In Lake Tahoe development struggle, California blinks, Nevada wins." *Los Angeles Times*, September 14, 2013.

Charmantier, A. and P. Gienapp. "Climate change and timing of avian breeding and migration: Evolutionary versus plastic changes." *Evolutionary Applications* 7 (2014): 15–28. doi:10.1111/eva.12126.

Curtis, John T. "The modification of mid-latitude grasslands and forests by man." In *Man's Role in Changing the Face of the Earth.* Chicago: University of Chicago Press, 1956.

Damschen, Ellen I., Nick M. Haddad, John L. Orrock, Joshua J. Tewksbury, and Douglas J. Levey. "Corridors increase plant species richness at large scales." *Science* 313, no. 5791 (2006): 1284–1286.

Directorate-General for Environment (European Commission). *Supporting the Implementation of Green Infrastructure, Final Report.* Publication: ENV.B.2/SER/2014/0012. Rotterdam: European Commission, May 31, 2016.

Dramstad, Wenche, James D. Olson, and Richard T. T. Forman. *Landscape Ecology Principles in Landscape Architecture and Land-Use Planning.* Washington, DC: Island Press, 1996.

Dunning, John B., Brent J. Danielson, and H. Ronald Pulliam. "Ecological processes that affect populations in complex landscapes." *Oikos* (1992): 169–1175.

Epps, Clinton W., Dale R. McCullough, John D. Wehausen, Vernon C. Bleich, and Jennifer Rechel. "Effects of climate change on population persistence of desert-dwelling mountain sheep in California." *Conservation Biology* 18, no. 1 (2004): 102–113.

Ernst, C. *Protecting the Source: Land Conservation and the Future of America's Drinking Water.* Washington, DC: The Trust for Public Land, 2004.

Ewers, Robert M., and Raphael K. Didham. "Confounding factors in the detection of species responses to habitat fragmentation." *Biological Reviews* 81, no. 1 (2006): 117–142.

Fahrig, Lenore. "Effects of habitat fragmentation on biodiversity." *Annual Review of Ecology, Evolution, and Systematics* 34, no. 1 (2003): 487–515.

Firehock, Karen. *Strategic Green Infrastructure Planning: A Multi-Scale Approach.* Washington, DC: Island Press, 2015.

Firehock, Karen E. *Evaluating and Conserving Green Infrastructure across the Landscape: A Practitioner's Guide.* Charlottesville, VA: The Green Infrastructure Center Inc., 2012.

Forman, Richard T. T., and Anna M. Hersperger. *Road Ecology and Road Density in Different Landscapes, with International Planning and Mitigation Solutions.* Publication No. FHWA-PD-96-041. Tallahassee: Florida Department of Transportation, 1996.

Forman, Richard T. T., and M. Godron. *Landscape Ecology.* New York: John Wiley & Sons, 1986.

Forman, R. T. T. *Land Mosaics: The Ecology of Landscapes and Regions.* Cambridge: Cambridge University Press, 1995.

Gates, J. Edward, and D. R. Evans. *Powerline Corridors: Their Role as Forest Interior Access Routes for Brown-Headed Cowbirds, a Brood Parasite of Neotropical Migrants.* No. PB–97-133516/XAB. Frostburg, MD: Appalachian Environmental Lab., University of Maryland, 1996.

Geertsema, Willemien. "Plant survival in dynamic habitat networks in agricultural landscapes." PhD diss., Wageningen University, 2002.

Gucinski, H., M. Furniss, R. Ziermer, and M. Brookes. *Forest Service Roads: A Synthesis of Scientific Information.* General Technical Report PNW-GTR-509.1. Portland, OR: USDA Forest Service, Pacific Northwest Research Station, 2001.

Haddad, Nick M., Lars A. Brudvig, Jean Clobert, Kendi F. Davies, Andrew Gonzalez, Robert D. Holt, Thomas E. Lovejoy, et al. "Habitat fragmentation and its lasting impact on Earth's ecosystems." *Science Advances* 1, no. 2 (2015): e1500052.

Hanski, Ilkka. "Predictive and practical metapopulation models: The incidence function approach." In *Spatial Ecology: The Role of Space in Population Dynamics and Interspecific Interactions*, edited by David Tilman and Peter Kareiva. Princeton, NJ: Princeton University Press, 1997.

Hanski, Ilkka, and Michael Gilpin. "Metapopulation dynamics: Brief history and conceptual domain." *Biological Journal of the Linnean Society* 42, no. 1–2 (1991): 3–16.

Harris, Larry D. *The Fragmented Forest: Island Biogeography Theory and the Preservation of Biotic Diversity.* Chicago: University of Chicago Press, 1984.

Harris, L. D. "Special visual presentation. Landscape linkages: The dispersal corridor approach to wildlife conservation." *Transactions of the 53rd North American Wildlife and Natural Resources Conference.* Louisville, KY, 1988.

Heller, Nicole E., and Erika S. Zavaleta. "Biodiversity management in the face of climate change: A review of 22 years of recommendations." *Biological Conservation* 142, no. 1 (2009): 14–32.

Hellmund, Paul Cawood, and Daniel Smith. *Designing Greenways: Sustainable Landscapes for Nature and People.* Washington, DC: Island Press, 2013.

Hengl, Tomislav. "Finding the right pixel size." *Computers & Geosciences* 32, no. 9 (2006): 1283–1298.

Hoctor, Thomas S., Margaret H. Carr, and Paul D. Zwick. "Identifying a linked reserve system using a regional landscape approach: The Florida ecological network." *Conservation Biology* 14, no. 4 (2000): 984–1000.

Huffaker, C. B. "Experimental studies on predation: Dispersion factors and predator–prey oscillations," *Hilgardia* 27, no. 14 (1958): 343–383.

Jenkins, C. N., K. S. van Houtan, S. L. Pimm, & J. O. Sexton. "US protected lands mismatch biodiversity priorities." *Proceedings of the National Academy of Sciences* 112, no. 16 (2015): 5081–5086.

Jennings, Michael D. "Gap analysis: Concepts, methods, and recent results." *Landscape Ecology* 15, no. 1 (2000): 5–20.

Jones, Claudia, Jim McCann, and Susan McConville. *A Guide to the Conservation of Forest Interior Dwelling Birds in the Chesapeake Bay Critical Area.* Annapolis, MD: Chesapeake Bay Critical Area Commission, 2000.

Kaplan, Rachel, and Stephen Kaplan. *The Experience of Nature: A Psychological Perspective.* Cambridge: Cambridge University Press, 1989.

Kapos, V., G. Ganade, E. Matsui, and R. L. Victoria. "∂^13 C as an indicator of edge effects in tropical rainforest reserves." *Journal of Ecology* (1993): 425–443.

Kilbane, Simon. "Green infrastructure: Planning a national green network for Australia." *Journal of Landscape Architecture* 8, no. 1 (2013): 64–73.

Krosby, Meade, Ian Breckheimer, D. John Pierce, Peter H. Singleton, Sonia A. Hall, Karl C. Halupka, William L. Gaines, et al. "Focal species and landscape 'naturalness' corridor models offer complementary approaches for connectivity conservation planning." *Landscape Ecology* 30, no. 10 (2015): 2121–2122.

Laurance, William F., Thomas E. Lovejoy, Heraldo L. Vasconcelos, Emilio M. Bruna, Raphael K. Didham, Philip C. Stouffer, Claude Gascon, Richard O. Bierregaard, Susan G. Laurance, and Erica Sampaio. "Ecosystem decay of Amazonian forest fragments: A 22-year investigation." *Conservation Biology* 16, no. 3 (2002): 605–618.

LAW 5/2014, of July 25, of the Spatial Planning, Urban Planning and Landscape, of the Autonomous Region of Valencia [2014/7303].

Lesk, Corey, Ethan Coffel, Anthony W. D'Amato, Kevin Dodds, and Radley Horton. "Threats to North American forests from Southern pine beetle with warming winters." *Nature Climate Change* 7, no. 10 (2017): 713.

Levin, Simon A. "The problem of pattern and scale in ecology: The Robert H. MacArthur Award Lecture." *Ecology* 73, no. 6 (1992): 1943–1967.

Levin, Simon A., and Robert T. Paine. "Disturbance, patch formation, and community structure." *Proceedings of the National Academy of Sciences* 71, no. 7 (1974): 2744–2747.

Levins, Richard. "Some demographic and genetic consequences of environmental heterogeneity for biological control." *Bulletin of the Entomological Society of America* 15, no. 3 (1969): 237–240.

Lewis, Renne, "Climate change played role in Everest avalanche, scientists say." *Al Jazeera*, April 26, 2014.

Lindenmayer, David B., and Jerry F. Franklin. *Conserving Forest Biodiversity: A Comprehensive Multiscaled Approach*. Washington, DC: Island Press, 2002.

Lines, Lee G., Jr., and Larry D. Harris. "Isolation of nature reserves in north Florida: Measuring linkage exposure." *Transactions of the 54th North American Wildlife and Natural Resources Conference*. Washington, DC, 1989.

Liquete, Camino, Stefan Kleeschulte, Gorm Dige, Joachim Maes, Bruna Grizzetti, Branislav Olah, and Grazia Zulian. "Mapping green infrastructure based on ecosystem services and ecological networks: A pan-European case study." *Environmental Science & Policy* 54 (2015): 268–280.

Little, Charles E. *Greenways for America*. Baltimore: Johns Hopkins University Press, 1995.

MacArthur, Robert H., and Edward O. Wilson. *The Theory of Island Biogeography*. Princeton, NJ: Princeton University Press, 2001.

McHarg, Ian L., and Lewis Mumford. *Design with Nature*. Garden City, NY: The Natural History Press for the American Museum of Natural History, 1969.

Mears and Wilber, "Identifying avalanche terrain."

Meffe, G. K., and C. R. Carroll. "Conservation reserves in heterogeneous landscapes." In *Principles of Conservation Biology*. 2nd ed. Sunderland, MA: Sinauer Associates, 1997.

Mladenoff, David J., Theodore A. Sickley, Robert G. Haight, and Adrian P. Wydeven. "A regional landscape analysis and prediction of favorable gray wolf habitat in the northern Great Lakes region." *Conservation Biology* 9, no. 2 (1995): 279–294.

Muñoz-Criado, A., and Vincente Domenech Gregori. *Comunitat Valenciana 2030: Síntesis de la Estrategia Territorial* [The Valencia Region in 2030: A Synthesis of the Regional Spatial Plan]. Valencia: Generalitat Valenciana, 2013.

Muñoz-Criado, A., et al. "La Nueva Política de Paisaje de la Comunitat Valenciana: Programas y Actuaciones 2011–2015" [The New Valencia Landscape Policy: Programs and Interventions, 2011–2015].

Muñoz-Criado, A., et al. "Guía metodológica de Estudios de Paisaje" [Methodology and Guidelines for Landscape Studies].

Muñoz-Criado, A. *Plan de la Huerta de Valencia. Un paisaje cultural milenario.* Vol. 1, *Estrategias de preservación y gestión.* Vol. 2, *Conclusiones del proceso de participación pública,* 2009.

Muñoz-Criado, Arancha, and Vicente Domenech. "Green Infrastructure Planning at Multiple Levels of Scale: Experiences from the Autonomous Region of Valencia, Spain." *Scale-sensitive Governance of the Environment* (2014): 283–301.

Noss, Reed. "Landscape connectivity: Different functions at different scales." In *Landscape Linkages and Biodiversity*, edited by W. E. Hudson. Washington, DC: Island Press, 1991.

Noss, Reed F. "The Wildlands Project: Land conservation strategy." *Wild Earth* (1992): 10–25.

Noss, Reed F., Michael O'Connell, and Dennis D. Murphy. *The Science of Conservation Planning: Habitat Conservation under the Endangered Species Act.* Washington, DC: Island Press, 1997.

Nussey, D. H., E. Postma, P. Gienapp, and M. E. Visser. "Selection on heritable phenotypic plasticity in a wild bird population." *Science* 310 (2005): 304–306.

Olson, David M., and Eric Dinerstein. "The Global 200: Priority ecoregions for global conservation." *Annals of the Missouri Botanical Garden* 89, no. 2 (2002): 199–224.

Parrish, Jeffrey D., David P. Braun, and Robert S. Unnasch. "Are we conserving what we say we are? Measuring ecological integrity within protected areas." *BioScience* 53, no. 9 (2003): 851–860.

Pulliam, H. Ronald. "Sources, sinks, and population regulation." *The American Naturalist* 132, no. 5 (1988): 652–661.

Robichaud, Isabelle, Marc-André Villard, and Craig S. Machtans. "Effects of forest regeneration on songbird movements in a managed forest landscape of Alberta, Canada." *Landscape Ecology* 17, no. 3 (2002): 247–262.

Russell, F. Leland, David B. Zippin, and Norma L. Fowler. "Effects of white-tailed deer (Odocoileus virginianus) on plants, plant populations and communities: A review." *The American Midland Naturalist* 146, no. 1 (2001): 1–26.

Rudolf-Miklau, Florian, Siegfried Sauermoser, and Arthur Mears, eds. *The Technical Avalanche Protection Handbook.* Imprint unknown, 2014.

Schröter, Matthias, Christian Albert, Alexandra Marques, Wolke Tobon, Sandra Lavorel, Joachim Maes, Claire Brown, Stefan Klotz, and Aletta Bonn. "National ecosystem assessments in Europe: A review." *BioScience* 66, no. 10 (2016): 813–828.

Scott, J. Michael, Frank Davis, Blair Csuti, Reed Noss, Bart Butterfield, Craig Groves, Hal Anderson, et al. "Gap analysis: A geographic approach to protection of biological diversity." *Wildlife Monographs* (1993): 3–41.

Schneider, David C. "The rise of the concept of scale in ecology: The concept of scale is evolving from verbal expression to quantitative expression." *BioScience* 51, no. 7 (2001): 545–553.

Söndgerath, Dagmar, and Boris Schröder. "Population dynamics and habitat connectivity affecting the spatial spread of populations—a simulation study." *Landscape Ecology* 17, no. 1 (2002): 57–70.

Sorrell, J. "Using geographic information systems to evaluate forest fragmentation and identify wildlife corridor opportunities in the Cataraqui watershed." Faculty of Environmental Studies, York University, Ont., Canada, 1997.

Steffan-Dewenter, Ingolf. "Importance of habitat area and landscape context for species richness of bees and wasps in fragmented orchard meadows." *Conservation Biology* 17, no. 4 (2003): 1036–1044.

Steinitz, Carl. *A Framework for Geodesign: Changing Geography by Design.* Redlands, CA: Esri, 2012.

Tansley, Arthur G. "The use and abuse of vegetational concepts and terms." *Ecology* 16, no. 3 (1935): 284–307.

Tewksbury, Joshua J., Douglas J. Levey, Nick M. Haddad, Sarah Sargent, John L. Orrock, Aimee Weldon, Brent J. Danielson, Jory Brinkerhoff, Ellen I. Damschen, and Patricia Townsend. "Corridors affect plants, animals, and their interactions in fragmented landscapes." *Proceedings of the National Academy of Sciences* 99, no. 20 (2002): 12923–12926.

Theobald, David M., Sarah E. Reed, Kenyon Fields, and Michael Soule. "Connecting natural landscapes using a landscape permeability model to prioritize conservation activities in the United States." *Conservation Letters* 5, no. 2 (2012): 123–133.

Thompson, Craig M., and Kevin McGarigal. "The influence of research scale on bald eagle habitat selection along the lower Hudson River, New York (USA)." *Landscape Ecology* 17, no. 6 (2002): 569–586.

Tilman, David, and Clarence L. Lehman. "Habitat destruction and species extinctions." *Spatial Ecology* (1997): 233–249.

Turner, Monica G., Robert H. Gardner, and Robert V. O'Neill. *Landscape Ecology in Theory and Practice.* Vol. 401. New York: Springer, 2001.

Turner, Tom. *Landscape Planning and Environmental Impact Design.* London: Routledge, 1998.

Ulrich, Roger, "View through a window may influence recovery." *Science* 224, no. 4647 (1984): 224–225.

van Dorp, Dirk, Peter Schippers, and Jan M. van Groenendael. "Migration rates of grassland plants along corridors in fragmented landscapes assessed with a cellular automation model." *Landscape Ecology* 12, no. 1 (1997): 39–50.

Von Humboldt, Alexander, and Aimé Bonpland. *Essay on the Geography of Plants*. Chicago: University of Chicago Press, 2010.

Walker, Richard, and Lance Craighead. "Analyzing wildlife movement corridors in Montana using GIS." *Proceedings of First Annual James Reserve Conference, July 5–7,* 1997.

Watt, Alex S., "Pattern and process in the plant community." *Journal of Ecology* 35, no. 1/2 (1947): 1–22.

Wells, Jeffrey V., and Milo E. Richmond. "Populations, metapopulations, and species populations: What are they and who should care?" *Wildlife Society Bulletin (1973–2006)* 23, no. 3 (1995): 458–462.

Whyte, William Hollingsworth. "Securing open space for urban America: Conservation easements." *Urban Land Institute* 2, no. 3 (1959).

Wiens, John A. "Population responses to patchy environments." *Annual Review of Ecology and Systematics* 7, no. 1 (1976): 81–120.

Wilson, Edward O., and Robert H. MacArthur. *The Theory of Island Biogeography*. Princeton, NJ: Princeton University Press, 1967.

With, Kimberly A., and Anthony W. King. "Dispersal success on fractal landscapes: A consequence of lacunarity thresholds." *Landscape Ecology* 14, no. 1 (1999): 73–82.

Wu, Jianguo, and Orie L. Loucks. "From balance of nature to hierarchical patch dynamics: A paradigm shift in ecology." *Quarterly Review of Biology* 70, no. 4 (1995): 439–466.

Wu, Jianguo, Weijun Shen, Weizhong Sun, and Paul T. Tueller. "Empirical patterns of the effects of changing scale on landscape metrics." *Landscape Ecology* 17, no. 8 (2002): 761–782.

Yahner, Richard H. "Changes in wildlife communities near edges." *Conservation Biology* 2, no. 4 (1988): 333–339.

Zipperer, Wayne C. "Deforestation patterns and their effects on forest patches." *Landscape Ecology* 8, no. 3 (1993): 177–184.

Index

303(d) list, 155–56
1000 Friends of Florida, 14–15, 213

A

AccuWeather, 1
acid rain, 40
Acquisition of Conservation Easements (ACE) program, 100
Adams Park scenario, 194–96; development scenarios, 194–96
advisory committees. *See also* experts: for determining opportunity, 173; in GI planning process, 23, 109–10; to identify needs, 99; importance of, 107; in least-cost path analysis, 117–18; for map-making, 106–7; in planning for connectivity, 121; for protecting scenic areas, 189; on ranking weights, 93; scientific, 107, 109–10, 118; technical, 93, 109–10, 117, 118; on web apps, 94
agriculture, 2, 40, 44, 48, 58, 70, 83, 85, 88, 96, 99, 100, 105, 130, 131, 133, 136, 144, 152, 200, 227
agritourism, 60
Aichi Biodiversity Targets, 20
air quality, 1, 2, 20, 24, 57
Albemarle County, Virginia, 100–102
American brown bears, 17
American pika, 183
Anza-Borrego Desert State Park, 13
apps. *See also* ArcGIS; ArcGIS tools/toolbox: to access, filter, and prioritize habitat cores, 91–95; for everyday planning, 204–9; to prioritize intact landscape cores, 92–93; to select intact landscape cores, 91–92
aquifers, 4, 27, 31, 41, 72, 82, 103, 147, 186, 191
ArcGIS: Desktop, 46, 93, 94; Feature Analyst software as extension of, 216; Living Atlas of the World, 97–98; metrics in, 30; Online, 45, 46, 91–93, 94; Pro, 45, 93, 118
ArcGIS tools/toolbox, 45, 88, 93; Combine tool, 90; Cost Connectivity tool, 118; Minimum Bounding Geometry tool, 108; in postprocessing, 216; Raster Calculator tool, 168; Reclassify tool, 148; Weighted Overlay tool, 149
Arizona Landscape Integrity and Wildlife Connectivity Assessment, 49
Arizona state model, 49

Arkansas Novaculite, 215
Arkansas state model, 16
Army Corps of Engineers, 168
asset maps, making, 103–39; basemap, 110–16; Carolina Bays, 108–9; compactness in, 108; experts and advisory groups on, 107, 109–10; formal advisory group, 107; for GI plan, 59–60; GIS lead, 106–7, 109–10; iterative process of, 106–7; multiple, interconnected benefits approach, 104–5; orientation in, 108; six-step process of, 59–60, 109; soil type in, 108; technical advisory committee, 109–10, 117
at-risk parcels, 191
attribute table, 30, 98, 108, 113, 132
Australia, GI network for, 224
avalanche zones, 166
Average Daily Trips (ADTs), 123

B

Balkan lynx, 223
basemaps, 86, 91, 125; creating, 110–16; data collection to build, 87–88; information included in, 25; intact habitat cores data, 88; mapping habitat cores, 110–12; ranking habitat cores, 112–15; in six-step process, 59–60, 68, 70, 78, 79; turning data for habitat cores into visual format, 115–16
bathtub model, 168
battlefield areas, 83
beaches, 198, 227
betweenness centrality (BC), 118–20
Big Bend region, 214
bighorn sheep, 17
binary approach, 148
Biodiversity Action Plan, 102
Biodiversity Conservation Strategy 2010–2020, 224
Biodiversity Priority Index, 113
biodiversity protection, 40, 92–93
biological information, adding to model, 123–24

BioMap2: Conserving the Biodiversity of Massachusetts in a Changing World, 49
Biosphere Conference, 12
biosphere reserves, 12–14, 223
brownfields, 181–82
buffer zones, 12–13, 15
burn-in process, 216

C

California Essential Habitat Connectivity Project, 49
California state model, 16, 49
canals, 104
Canyon De Chelly National Monument, Arizona, 3
Capital Region Land Conservancy, 209–10
Carolina Bays, 108–9
Carr, Margaret, 47
cartography tips for GI maps, 133–35; classification schemes for quantitative GIS data, 134; colors and symbols, 133–34; representing data, 134; topography, 134–35; unnecessary detail, 134
case studies, 209–27; European Green Belt initiative, 222–24; Florida Wildlife Corridor, 212–14; informing land trust conservation priorities, 209–11; urban GI in Hot Springs, Arkansas, 214–22; Valencia, Spain, GI planning leadership in, 224–27
Changbai Mountains, 13–14
change drivers, 152, 200
Charlottesville, Virginia, 104, 181
Cherokee National Forest, 179
Chesapeake Bay Program, 158–59
Chiles, Lawton, 15
Clark Labs, 152–54
classification schemes for quantitative GIS data, 134
clean air, 1, 2, 20, 24, 57
clean water, 2, 4, 7–8, 20, 38, 62, 155, 175
Clean Water Act, 155
climate change, 2–3, 163–64; avalanches and, 166; habitat conditions affected by, 1, 200; mitigation techniques, 34; natural stressors

and, 141–42; opportunities, 196–201; pest outbreaks and, 161; regional modeling and, 46; resilient areas and, 211; risks and, 163–64; scale and, 40–41, 46; sea level rise and, 167, 196–99; sequestering carbon to mitigate, 71; water quality and, 196

Coastal Change Analysis Program (C-CAP), 90

code suitability, 133

Colorado state model, 16

color scheme across maps, 133–34

Community Rating System (CRS), 208

community support for GI plan, 72–78; criteria to select cores, 72–75; ensuring, 72–75; gaining, 76–78; goals, examples of, 72

compactness ratio, 108, 113

composite ranking, 114, 115

connectivity across landscape, modeling, 38–43; cost-distance analysis, 39–40; extent and resolution, determining, 42–43; geographic extent, 41; scale, concepts of, 40–41; spatial resolution, 41

connectivity planning: factors to consider, 120–21; questions to help prioritize connections, 121–23; tools to help inform, 118–20

Connectors dataset, 117, 118, 120–21

conservation, choices we make for, 64

conservation easements, 16, 35, 62, 73, 100, 121, 142–43, 144, 153, 175, 183–84, 204, 206, 209–11

Conservation Fund, 14, 16, 18, 206, 208

conservation plans, informing strategic, 207

Conservation Science Partners, 18

constraints of time, staff resources, and budget, 148–51; binary approach, 148; weighted overlay approach, 148–51

contagion, 38

contamination: Chesapeake Bay Program, 158–60; contaminated lands, 160–61; evaluate by known, 155–61; impaired (polluted) waters, 155–60

contrast, 38

Convention on Biological Diversity (CBD), 19–20, 224

cores, 11, 24. See also habitat cores and corridors; intact habitat cores data; landscape corridors; mapping connectivity between cores; community values to determine, 135–36; criteria to select, 72–75; data by state, download for, 45; edge zone, 30; flag, 174; fragmentation of, in rural-urban interface, 138–39; interior zone, 30; map, 124–25; origin of term, 12; prioritizing, based on topical goals, 136; ranking, 114, 116, 162; reranking, 136–37; size of, in acres, 113; strategic location of, in meeting goals, 182–90; thickness of, 113; two-part, notion of, 30

corridors: defined, 34; habitat, 11; importance of, 24

cost-distance analysis, 39–40, 44

cost rasters, 40, 44

cost surface, 39–40, 44, 89, 118

critical linkages, 212

cross-county connectors, 183–84

cultural assets, 55, 83, 97, 103–5, 181, 207

cultural resources, 128

cultural sites, 60

culture themed overlays, 130, 137

D

Dana Biosphere Reserve, 13, 14

Darlington County, South Carolina, 73–74, 108

data. See also intact habitat cores data: collecting, reasons for, 81, 84, 85–88; gaps, filling, 96–98; habitat cores to assess zoning changes, 100–102; local, adding, 95–99; mapping and, 98–99; representing effectively, 134; reviewing, in six-step process, 59, 109; scales, 98–99; surrogates, 98

death by a thousand cuts, 43

Death Valley National Park, 13

decision metrics, 76

decluttering map, 134

Defenders of Wildlife, 16, 213
deforestation, 43
Department of Forestry, 83, 85
detail of maps, avoiding unnecessary, 134
development costs, 206
development pressure overlay, performing, 146–47
development scenarios, 190–96; Adams Park scenario, 194–96; alternative, 190–93
digital elevation model (DEM), 90, 134
disrupting features, risk evaluated by, 164–65; pipelines and mines, 164–65
ditches, 124
drains, 124
dump sites, 160–61, 181–82
dunes, 25, 57, 84, 147, 167, 198, 227

E

Earth Observatory, NASA, 170
Eco-logical, 181
ecological characteristics, in cores dataset, 111
Ecological Connectivity Project, GNLCC, 46
ecological core model data, 131
ecological integrity, 27, 30–31, 45–46, 50, 68, 72, 90, 112
ecological land unit diversity, 113
Ecological Land Units (ELUs), 90
ecologically relevant geophysical (ERGo) landforms, 90
ecological system redundancy, 90, 113
economy, 2
ecoregional assessments, 89
ecoregion ranking, 114
ecosystem, 26
ecosystem services, 57
ecosystem services values, 20
ecotourism, 65
edges, 31–33
edge zone, 30
element occurrence (EO), 123, 124
endangered species, 4, 5, 10, 14, 20, 27, 41, 73, 76, 97, 99, 124, 142–43, 177, 208

Endangered Species Act (ESA), 10, 235
endemic species max, 114
environmental protection, 2, 24
Environmental Protection Agency (EPA), 155
EO Reps, 123
equal interval classification scheme, 134
Esri Guide to GIS Analysis, Volume 3: Modeling Suitability, Movement, and Interaction, 40
Esri Habitat Cores, 99, 113
Esri SSURGO downloaded web app, 131
Eurasian otter, 223
European Centre for Nature Conservation, 20
European Commission (EC), 20–21
European Green Belt initiative, 222–24
European Union (EU) countries, 19–21, 224
Everglades National Park, 214
experts. *See also* advisory committees: to build support for goals, 78; data gaps and, 97; for determining opportunity, 173; Esri cores model and, 23, 75–76; GIS lead seeking expertise from, 106–7; habitat cores model and, 23; to identify mapping issues and suggest refinements, 75; on review committee, 110; risks revealed by, 170; on sea level rise, 169; on water recharge areas, 186; working with, 107

F

Feature Analyst software, 216
Federal Emergency Management Agency (FEMA), 199
fire-risk map, 3, 162
fires, 1, 3, 8, 57, 72, 138, 141, 142, 162–63, 170, 235
fisheries, 1, 58, 69, 70, 82, 156, 200, 214
flag cores, 174
flood-prone area, 3, 67
floods, 1, 8, 141, 215, 235
Florida Ecological Greenways Network (FEGN), 15, 47, 212–13
Florida Forever fund, 212
Florida Greenways Commission, 14–15
Florida state model, 47

Florida Wildlife Corridor, 212–14
focal species, 29–30
forest edges, 31–32, 138
forestry, 132; agency, 71, 83, 85, 175; land-use designation, 152; management, 83, 130, 133; mapping, 99, 130; parcel data, 131, 132; production, 38, 131; tax relief, 96; tracts, 60, 62, 78
Forest Stewardship Program's Spatial Analysis Project, 133
fragmentation, 27–29, 43, 138–39
FRAGSTATS 4, 28

G

Gap Analysis Program (GAP), 90, 113
GAP Land Cover maps, 90
General Conference, UNESCO, 12–13
geographic extent, 41
geographic information systems (GIS), 3; lead, 106–10, 133, 146, 173; software, 23 (see also ArcGIS); toolbox, 45
geology, 8, 26, 114, 186, 215
geometry, in cores dataset, 111
ghost subdivisions, 145, 190
GIS lead, 106–10, 133, 146, 173
Glades to Gulf Expedition, 214
Global Land Cover Facility, 20
goals: community support for, 72–78; decision metrics, 76; recreation goals, 187–88; sample goals for GI plan, 69–71; scenic goals, 188–90; six-step process for setting, 57–59, 65–80; stakeholder engagement, 72, 77; strategic location of cores in meeting, 182–90; themed overlays for, 124–35; time frame and specificity for, 75–76; water quality goals, 184–85; water supply goals, 186–87; wildlife goals, 183–84
grain of the data, 41
gray infrastructure, 24, 86
Grayson County in Southside Virginia, 124

Great Northern Landscape Conservation Cooperative (GNLCC), 46
green infrastructure (GI): basemap, 25, 86, 91, 125; modeling, 23–50; network, 8–14; purpose of, 2
Green Infrastructure Center (GIC), 16, 18, 19, 23, 48, 49, 55, 56, 87, 102, 195, 206, 210, 214, 220, 227
Green Infrastructure (GI)—Enhancing Europe's Natural Capital, 222–24
green infrastructure (GI) modeling, 23–50. *See also* connectivity across landscape, modeling; of accumulated changes, 43–44; biological information, adding to model, 123–24; building, 25–38; composition, measures of, 38; cores model for identifying landscape areas as habitat cores, 75–76; defined, 25; ecological core model data, 131; experts and, 23, 75–76; extent to which natural lands are connected across, 29; fragmentation, 27–29; habitat data, 87; healthy community and, 6–7; intact habitat core data for nation, 93; intactness, 27–29; of large landscapes, 43; metrics, 37–38; national model, 18, 44–50, 87–88, 210; nationwide modeling, 44–46; permeability, 44; planning, 42, 95; purpose of, 24–25; regional modeling, 46; state models, 47–50; technical terms, other, 38; terminology, 25–26; tourist industry and, 7; in US, 14–18
Green Infrastructure Network Model for South Carolina, 211
green infrastructure (GI) planning. *See also* implementing GI plans: around the world, 18–21; for clean watershed, 7–8; community support for, 72–78; ecology of, 44–45; for healthy community, 6–7; key terminology in GI network design, 8–14; landscape elements central to, 29–37; maps for, 5; new way to plan, need for, 5–8; sample goals for, 69–71; for tourist industry, 7; in urban areas, 207
Green Infrastructure Regional Plan, 225–26

GreenPrint, Maryland's, 48
Green Team, 99–100
greenway design, 236–37
Greenways Commission, 14–16
grizzly bears, 17
groundwater recharge areas, 3, 31, 58, 60, 69, 82, 212
Gulf Islands National Seashore, 214

H

habitat, 26
habitat cores and corridors, 11. *See also* cores; intact habitat cores data; assessing risks to, 61; to assess zoning changes, 100–102; Esri cores model for identifying landscape areas as, 75–76; intact, data for nation, 88–95; mapping, 110–12; ranking, 112–15; turning data into visual format for, 115–16
habitat fragmentation, 10, 27–29, 43, 59, 88–89, 121–23
hazard mitigation, 67
health care costs, 2, 24
healthy lifestyles, 2, 4, 221
heat map, 147, 149, 152
heritage themed overlays, 130, 137
high biological component weighting, 114
high-tide event, 170
high-tide flooding, 170
hiking trails, 31, 58
hillshade, 134–35
Hoctor, Tom, 47
Hot Springs, Arkansas, 99–100, 214–22; example strategies, 220, 222; land cover classification, 215–16; postprocessing, 216–20
hub. *See* cores
Hurricane Harvey, 1, 170
hydrology, 1, 8, 20, 26, 81

I

impaired landscapes, 181–82

impaired (polluted) waters, 155–60; Chesapeake Bay Program, 158–60; list, 155–56
implementation, defined, 222
implementing GI plans, 203–28; case studies, 209–27; decision support, 204–9; informing strategic conservation plans, 207; natural asset maps, 204; projects initiated by other public bodies, 205–6; regulatory needs, meeting, 207–9; saving money by using GI maps, 206; six-step process, 62–63; stovepipe planning, overcoming, 205
independent parcels, opportunities at level of, 175–76
intact habitat cores data, 88–95; accessing, 88; apps to access, filter, and prioritize habitat cores, 91–95; core size used in, 31; downloading, 88–94; gaps in, filling, 96–97; limitations of, 95; local, adding, 95–96; map packages, 88–91; metrics used in, 111–14; model, 93; ranking habitat cores, 112–15; scales, 98–99; summary scores calculated for, 113–14; surrogates, 98; toolbox, 94; turning into visual format, 115–16; use cases, 94; values assigned to, 114
interior zone, 30
interspersion, 37, 38
invasive species, 30, 31–32, 35, 41, 138, 161, 181, 197
isolation, 28

J

Jaguar Corridor Initiative, 19
Jefferson, Thomas, 104, 189
Jefferson County, Colorado, 111–12, 115, 116, 161, 162
joint management areas, 13–14
Joshua Tree National Park, 13

K

Karner blue butterfly, 17

kernel density estimation, 152
king tide, 170

L

lakes, 2, 27, 69, 70, 71, 89, 117, 155, 197, 215
Lake Tahoe, 153–54
land cover types, 37, 38, 107, 134, 216
landfills, 181–82
Landsat Vegetation Continuous Fields tree cover layer, 20
Landscape Conservation Cooperatives (LCCs), 46
landscape corridors, 117–24. See also mapping connectivity between cores
landscapes, impaired, 181–82
landscape suitability, 133
landslide models. See green infrastructure (GI) modeling
landslide risk map, 166
landslides, 166
land trust conservation priorities, informing, 209–11
Land Use Conflict Identification Strategy (LUCIS), 149
land-use designation, 152
land-use planning, 23, 35, 46
larger-scale opportunities, 176–77
leapfrogging subdivisions, 144
least-cost path analysis, 40, 46, 89, 117–18, 120–21
Legacy Institute for Nature & Culture (LINC), 213
LiDAR-derived DEM, 168
lifestyles, healthy, 2, 4, 221
life-sustaining services, 2, 24, 27, 57
livability, 221, 222
local knowledge in map-making process, 66–67; hazard mitigation, 67; value of, 67

M

Magnolia Plantation and Gardens, 104
Malvern Avenue Gateway, 222
management opportunity, 133
management overlays, 130
Man and Biosphere (MAB) Program, 12
manual interval classification scheme, 134
map-making process, 67–69. See also asset maps, making; six-step process
mapping connectivity between cores, 117–24. See also landscape corridors; adding biological information to model, 123–24; factors to consider when planning for connectivity, 120–21; least-cost path analysis, 117–18; questions to help prioritize connections, 121–23; tools to help inform connectivity planning, 118–20
maps. See also cartography tips for GI maps: development pressure overlay, performing, 146–47; evaluating risks, 141–42; for everyday planning, 204–9; natural asset maps, 204; packages, 45, 88–91, 118, 131; risk-assessment map, items added to, 146; risks not easily revealed by, 170–71; rules, 30, 73, 76, 109, 123, 148; saving money by using, 206; using, 5
Maryland GreenPrint, 16, 48
Maryland Greenways Commission, 16
Maryland state model, 48
Massachusetts state model, 49
Master Address File (MAF/TIGER) database, 89
McHarg, Ian, 1
metapopulation, 36, 235–36
metrics: decision, 76; in GI modeling, 37–38; in intact habitat cores data, 111–14; mapping, 112
mines, 164–65; reclaiming old mine sites, 165
Minnesota Central Lakes region, 117
modeling. See green infrastructure (GI) modeling
Mojave Desert, 13
Montana Connectivity Project, 49, 50
Montana state model, 16, 49, 50
Monticello, 104, 189
mosaic, 8, 9, 37, 38, 234, 235
mountain lions, 31
mountain pine beetle, 161
mule deer, 34
multibenefit thinking, 199
multiple, interconnected benefits approach, 104–5

Muñoz-Criado, Arancha, 225

N

NASA, 170
National Agriculture Imagery Program (NAIP), 215
National Ecological Networks of European Countries Map, 20
National Elevation Dataset (NED), 90
National Flood Insurance Program (NFIP), 208
National Hydrography Dataset (NHD) Plus modifier, 89, 114
National Land Cover Database (NLCD), 25, 42, 45, 89, 90, 93, 94, 95
national model, 44–50; adaptable, 45–46; creation of, 18, 44, 87–88; data, accessing, 45; data used to create, 18; ease in using, 46; locally relevant, 45; regional modeling, 46; South Carolina model and, 49, 210; state models, 47–50; tools used to increase accuracy or specificity of, 50
National Oceanic and Atmospheric Administration (NOAA), 85, 90, 168
National Park Service, 100, 215
national ranking, 114
National Register of Historic Places, 106
National Wetlands Inventory (NWI), 70, 81, 89, 130, 197
National Wildlife Corridors Plan, 224
nationwide modeling, 44
Natura 2000 initiative, 21
natural asset maps, 62, 83, 180, 204
natural breaks classification scheme, 134
natural hazards, risk evaluated by, 165–70; avalanche zones, 166; landslides, 166; sea level rise, 167–70
Natural Heritage Program, 48
natural landscape-based themed overlays, 130, 136
naturalness, 44
nature-based recreational areas, 60, 129, 136
Nature Conservancy, The (TNC), 14, 16, 89, 101, 206, 211, 213

NatureServe, 16, 18, 90
New York state model, 16
nonforested areas, 8, 179
nongovernmental organizations (NGOs), 18, 46, 223
normalized difference vegetation index (NDVI), 216
Northwoods area, 219, 221, 222
nuisance flooding, 170

O

off-the-shelf data, 87
Okefenokee National Wildlife Refuge, 214
Olmsted, Frederick Law, 225
on-the-ground conservation work, 21
opportunities: Adams Park scenario, 194–96; climate change, 196–201; determining, 173–201; development scenarios, 190–96; impaired landscapes, 181–82; larger-scale, 176–77; at level of independent parcels, 175–76; multibenefit thinking, 199; prioritizing, 173–81; rarity, assessing, 177–78; restoration potential, assessing, 178–80; in six-step process, 62; strategic location of cores in meeting goals, 182–90; transportation plans as, 180–81
orientation in asset maps, 108
outdoor recreation-based themed overlays, 130, 136
overlays, 86. *See also* themed overlays; benefits of, 5; in GI planning, 20, 39
overlay zoning, 176
overlooks, 189, 207

P

pan-European Green Infrastructure Network, 20–21
parcel data, 98, 130, 132
parks, 2, 5, 16, 25, 39, 47, 60, 63, 65, 66, 70, 86, 95, 99–100, 125, 130, 133–34, 136, 142, 143, 146, 147, 182, 204, 205, 206, 207, 215, 223, 226, 227

Patch Analyst 5, 28
patch dynamics, 8, 234, 236
patches, 8, 11, 24; mosaic of, 8, 9, 37, 38, 234, 235
paths of least resistance, 44, 89
Pee Dee Land Trust (PDLT), 210–11
Perkl, Ryan M., 64
physical characteristics, in Esri cores dataset, 111
Piedmont region of central Virginia, 100, 181
pipelines, 164–65
Planning District Commission (PDC), 210
pollution-loading reductions, 195–96
polygons, 44, 88, 89, 108, 152
ponds, 70, 130
prescribed burns, 163
projects initiated by other public bodies, 205–6
project teams, 106
pronghorns, 34, 46
proxy data, 98

Q

quality of life, 1, 2, 24
quantile classification schemes, 134

R

ranking cores, 114, 116, 162
rarity, assessing, 177–78
recreation: goals, 187–88; nature-based recreational areas, 60, 129, 136; outdoor recreation-based themed overlays, 130, 136
recreationists, 2
regional modeling, 46
regulatory needs, meeting, 207–9; National Flood Insurance Program's Community Rating System, 208; state wildlife planning, 208–9
reranking cores to meet new priorities, 135–37; based on topical goals, 136–37; community values to determine, 135–36
reservoirs, 7, 82, 126, 130
Resilience Network Model (RNM), 211
restoration potential, assessing, 178–80

review committee, 110
Richmond Regional Planning District Commission (PDC), 210
risk-assessment map, items added to, 146
risks: to assets, assessing, 141–71; climate change, 163–64; constraints of time, staff resources, and budget, 148–51; fire risk and prescribed burns, 162–63; natural hazards, 165–70; not easily revealed by maps, 170–71; pest outbreaks and climate change, 161; in six-step process, assessing, 60–61, 110; types of, mapping and evaluating, 141–42; zoning ordinances and other protective measures to provide safety, 63
risks, evaluating: by disrupting features, 164–65; future change based on past trends, 152–55; by known contamination, 155–61; by level of pressure and proximity of development, 144–71; by level of protection, 142–44; limitations of, 147; types of risks, 141–42; by zoning or land-use designation, 152
rivers, 2, 7, 16, 47, 60, 71, 83, 89, 123, 126, 130, 167, 180, 186, 197, 200, 207
roads data, 87, 94, 117
Roosevelt, Theodore, 3
ruffed grouse, 180
Ruffed Grouse Society, 180
rural-urban interface, 138–39

S

sage grouse, 10
salt meadows, 223
Santa Rosa and San Jacinto Mountains National Monument, 13
scale: concepts of, 40–41; ideal, for all species, 42; ideal, for assessing habitat fragmentation, 43
scenario building, 194–96
scenic goals, 188–90
scenic rivers, 83
scenic views, 66, 136, 188, 224
scientific advisory committees, 107, 109–10, 118

sea level rise (SLR), 167–70, 196, 197, 199–200
Shannon-Weaver index, 113, 243–44, 248
six-step process, 55–80; overview of process, 55–57; overview of steps, 57; Step 1: Set goals, 57–59, 65–80; Step 2: Review data, 59, 109; Step 3: Make asset maps, 59–60, 109; Step 4: Assess risks, 60–61, 110; Step 5: Determine opportunities, 62; Step 6: Implement a plan, 62–63
sky islands, 183
smart growth, 2, 191
Soil Survey Geographic (SSURGO) database, 89, 108, 130, 131
soil type in asset maps, 108
Sonoran Desert tortoise, 183
South Carolina state model, 16
Southern Forest Landscape Assessment (SFLA), 133
spatial configurations, 38, 236
spatially complex information, 134
spatial resolution, 41
species: endangered or at-risk, 20; endemic species max, 114; focal, 29–30; ideal scale for all, 42; invasive, 30, 31–32, 35, 41, 138, 161, 181, 197; isolating populations of, risk of, 17; in Minnesota Central Lakes region, 117
Species of Greatest Conservation Need, 208
sprawl, 17
stakeholders in GI planning: in multiple countries, 20, 21, 224; regional modeling, 46; in review committee, 110; roles of, 72; Ruffed Grouse Society, 180; to support goals, 77, 78, 80
standard deviation, 113, 134; classification scheme, 134
state map packages, 118
state models, 16, 47–50; Arizona, 49; Arkansas, 16; California, 16, 49; Colorado, 16; Florida, 47; Maryland, 48; Massachusetts, 49; Montana, 16, 49, 50; New York, 16; South Carolina, 16; Virginia, 16, 48–49; Washington, 49; West Virginia, 16
state wildlife planning, 208–9

stepping stones, 36–37, 38, 50, 62, 89, 117, 121–22, 178–79, 182, 207
stovepipe planning, overcoming, 205
Strategic Regional Plan for the Region of Valencia 2030, 226
stream density, 75–76, 92, 113
sunlight, 31–32, 180
superfund sites, 160
Surface Mining Control and Reclamation Act (SMCRA), 165
symbols across maps, 133–34

T

Targeted Ecological Areas (TEAs), 16
technical advisory committees, 93, 109–10, 117, 118
Textron Systems, 216
themed overlays: common, 124; cores map, 124–25; creating to meet goals, 124–35; cultural resources, 128; culture or heritage, 130, 137; data to guide action, 126–29; examples of, 60; forestry, 132; habitat map, 125; nature-based recreation, 129; outdoor (natural landscape-based) recreation, 130, 136; reranking cores to meet new priorities from, 135–37; to show new priorities, 125, 131; in six-step process, 59–60; SSURGO Downloader web app and, 131; in state models, 48; water (elements and uses), 130, 136; water quality and watershed protection, 126–27; what to include, 130; working lands theme, 130–32, 136
Theobald Human Modification Index, 90
threats. *See* risks
topographic diversity, 48, 111, 113
topography to enhance maps, 134–35
Topologically Integrated Geographic Encoding (TIGER), 89
total maximum daily loading (TMDL), 156, 158–59
tourism, 2, 7, 39, 58, 60, 63, 72, 78, 100, 124, 125, 128, 130, 183, 199, 221
transportation plans as opportunities, 180–81

tree canopy, 60, 71, 84, 122, 207, 214–18
tributaries, 130
Trust for Public Land, The, 16, 18

U

underlay, 86
United Nations Educational, Scientific and Cultural Organization (UNESCO), 12–13
United States (US): GI modeling in, 14–18; sprawl in, 17
US Department of Transportation, 181
US Geological Survey (USGS), 163, 166
"US Protected Lands Mismatch Biodiversity Priorities," 113
urban areas, GI planning in, 207. *See also* Hot Springs, Arkansas
urban tree canopy (UTC), 84
USGS Sea Level Rise Modeling Handbook, 168

V

Valencia, Spain, 224–27
Valencia Regional GI plan, 226
Vaux, Calvert, 18
viewsheds, 61, 71, 83, 86, 130, 133, 188–89
Virginia Department of Conservation, 48–49
Virginia Natural Landscape Assessment, 48–49
Virginia state model, 16, 48–49
von Humboldt, Alexander, 18

W

Washington Connected Landscapes Project, 49
Washington state model, 49
water infrastructure, 130
water quality: climate change and, 196; goals, 184–85; NAIP imagery and, 216; of stormwater, 186
water recharge areas, 186–87
water resources, 48, 81–83, 105
watershed protection, 126–27

water supply, 1, 58, 62, 69, 71, 79, 99, 124, 135, 164, 173, 175, 181, 203, 204; goals, 186–87; water recharge areas and, 130, 186–87, 206
water (elements and uses) themed overlays, 130, 136
Watt, Alex, 8, 234
weighted overlay approach, 148–51
West Virginia state model, 16
wetlands, 8, 25, 59, 60, 61, 69–70, 81, 84, 89, 90, 96, 108, 113, 126, 130, 133, 147, 151, 180, 182, 200, 223, 235
Wilderness Society, The, 16
wildland–urban interface, 138
wildlife goals, 183–84
Wildlife Habitat Conservation Areas, 138
wildlife management areas (WMA), 70, 82, 130, 136, 142
wildlife planning, state, 208–9
wildlife-sustaining habitat cores, 24
wind, 8, 31–32, 170, 235
wolves, 28
working landscapes, 16, 47, 60, 83, 85, 124, 125, 130–31, 133, 136
working lands theme, 130–32, 136; code suitability, 133; ecological core model data, 131; Esri SSURGO downloaded web app, 131; forestry, 132; landscape suitability, 133; management opportunity, 133

Y

Yellowstone National Park, 17, 236

Z

Zig-Zag Mountains, 215
zoning: changes, habitat cores to assess, 100–102; evaluating risks by, 152; ordinances and other protective measures to provide safety, 63; overlays, 130
Zwick, Paul, 47